DO
UNTO
Animals

DO
UNTO
Animals

A FRIENDLY GUIDE TO
How Animals Live,
and How We Can Make
Their Lives Better

Tracey Stewart

ILLUSTRATIONS BY LISEL ASHLOCK

ARTISAN
New York

Library of Congress Cataloging-in-Publication Data
Stewart, Tracey
  Do unto animals : a friendly guide to how animals live, and how we can make their lives better / Tracey Stewart ; illustrated by Lisel Ashlock.
    pages cm
  Includes bibliographical references and index.
  ISBN 978-1-57965-623-2 (alk. paper)
  1. Human-animal relationships. 2. Animals—Habitations. 3. Animal communication. 4. Pets. 5. Livestock. I. Ashlock, Lisel Jane. II. Title.
  QL85.S793 2015
  590—dc23
    2015010995

Artisan books are available at special discounts when purchased in bulk for premiums and sales promotions as well as for fund-raising or educational use. Special editions or book excerpts also can be created to specification. For details, contact the Special Sales Director at the address below, or send an e-mail to specialmarkets@workman.com.

Published by Artisan
A division of Workman Publishing Company, Inc.
225 Varick Street
New York, NY 10014-4381
artisanbooks.com

Published simultaneously in Canada by Thomas Allen & Son, Limited

Printed in Malaysia
First printing, September 2015

10 9 8 7 6 5 4 3 2 1

To Jon, Nate, and Maggie

Thank you for giving me to this project
many, many, many times.
For inspiring and supporting me.
For sharing in the joy, the love, the care,
and the respect for our very large family.

Contents

Introduction

LOVE OF ANIMALS was part of me from the very beginning. There are pictures of my mother pregnant with me, a bird on her head, a rabbit in her arms, and a dog at her feet. It was the late 1960s, so also by her side sat Aunt Ginny puffing away on a Pall Mall cigarette. I like to think I was soaking up more of the animal pheromones than of the secondhand smoke, but it makes for a convenient excuse for any deficiency I may have. As much as I loved animals, though, I never thought I could actually pursue a career working with them . . . for a couple of reasons.

1. I was not good at math. I remember telling Mrs. Jessup, my second-grade math teacher, that my dream was to become a veterinarian when I grew up. She quickly admonished me with "You'll never be a veterinarian unless you apply yourself more in math." I heeded her words and decided then and there that I would do something that had nothing to do with math.

2. I came from a long line of scrappers. Each generation of my English/Irish family had inched their way up from a baseline of poverty, bettering themselves a little bit more than the previous generation but always being mindful to not be considered by the elders as getting too big for their breeches. We were an incredibly kindhearted crew with a confounding mix of arrogance and insecurity, lovable and deeply flawed. We were not a group that followed our passions, despite the fact that we were quite passionate. Instead we figured out a safe and practical path: a good job that we may not have particularly liked but that by God we would do diligently until the sweet relief of weekends, holidays, or death! Thinking I would become a veterinarian would have been too lofty, and working with animals in any other way would have been considered folly.

But I was good at art. It came easily to me, so my career path was decided. I went to a college that offered not only art but also business . . . safe. I spent my twenties and thirties working in every area a design degree might take you: art gallery, waitressing, architecture, interior design, graphic design, bathing suit and lingerie design. I've never not gotten a job I interviewed for and I've never been fired, which means I had a lot of jobs that I didn't actually like and I stayed in those jobs for far too long. From the outside, it probably looked like I loved all my jobs. I was a hard worker. In reality I dreaded every weekday morning and was exhilarated to be released at the end of each workday.

Luckily for me, I met my future husband and he quickly went to work studying me. He was confused that I was seemingly so capable of great joy and emotion and yet spent a good portion of my life bored to tears and un-inspired. It was the first time someone had called into question the risk of taking the safe route. Of course, I didn't go down without a fight. There were other rationalizations I had invented to keep me comfortably bored. Animals always occupied a lot of space in my heart. If I were to work with animals on a more regular basis, how would I ever be able to handle the sheer volume of loss I perceived I would experience? When I was nine, I signed up to receive newsletters from an animal rights organization. My parents had to intercept the mailers before they got to me because seeing the graphic images of animals suffering was something I couldn't easily recover from. In contemplating a transition into the veterinary field, I theorized that I wouldn't be able to see an animal in pain, and I certainly wouldn't be able to handle death or euthanasia. My boring job that didn't make me cry held a strange comfort for me. My man challenged me, though. He believed that I was the kind of person who needed to have a job that made me cry. He was right. (Note: This is not to say that my husband is always right, because more often than not I am the one who is right, but I'll give him this one.)

I decided to go back to school for a degree in veterinary technology. Joy and inspiration were what I found. Working in a veterinary clinic brought many tears, but most were happy. Those photos of animals suffering that upset me when I was young did so partially because they made me feel helpless, and now I was no longer helpless. I was prepared and right there, ready to provide care. That person who dreaded death and euthanasia became the go-to tech for those moments. I found it profoundly rewarding to be able to comfort both the animals about to pass and the people who would miss them beyond words when they were gone. I witnessed so many beautiful moments during those times. Another surprising thing that happened was that I became good at math. Apparently, it just needed to be applied to something I cared about. Take that, Mrs. Jessup!

My flight on animal love endorphins took a four-year detour when I had my kids. Caring for them when they were infants and toddlers felt quite similar to my work in veterinary medicine. Beautiful innocence, constant checking of vitals, and lots and lots of poop. During this time I created a café that focused primarily on feeding exhausted, overwhelmed parents with nourishing organic fare while engaging kids in their natural love of animals through classes and art projects. Parents were being nurtured up front while their kids were learning about compassion in the back. It was lovely while it lasted, but as my kids started to get older they were begging to get out into the world. I knew from experience the gift that animals offered me, and I was eager to share that with my children in a more meaningful way. I was also happy to get out of the café's basement and away from payroll. Just because I didn't fear math anymore didn't mean I liked it.

Luckily, my kids are as obsessed with animals as I am. We began volunteering at a local shelter as a family. My kids would read to all the animals and make adoption videos of the dogs and cats that were especially good with kids. Rather than feeling powerless to help as I did when I was a kid, my own kids felt empowered. They were able to be advocates, defenders, and nurturers. Their voices were heard and they made things happen. In two instances specifically, a shelter dog was adopted the day after their video aired. Their creativity and ingenuity were on fire. They came up with animated adoption videos, bake sales, jewelry sales, craft sales, and any other type of sale to raise awareness and funds for their now close friends at the shelter. They gave up getting birthday presents, instead collecting shelter supplies. They told friends, relatives, classmates, and anyone who would listen about the potential great buddy that was just waiting for them at the shelter. When a dog or cat friend got adopted and it was time to say good-bye, there were tears—but I now knew how essential crying was. That taste of selflessness was much needed for their little developing brains and hearts.

If you're an animal lover of any age, there is no end to the easy and fun ways you can bring animals into your life. While giving of yourself, you will most likely find you're getting a lot more back in return. Your home, backyard, local shelter, and neighboring farm sanctuaries are just some of the places that provide opportunities. When it comes to kids and animals, you really don't have to wait for someone to develop a program for them to get involved in. Kids hold a lucky secret pass to lots of things. On a recent vacation, we went into a shelter we'd never visited before. My daughter walked up to the receptionist, and though her face couldn't be seen over the desk, her voice could be heard saying, "I brought a book, *A Tale of Two Guinea Pigs*. Can I read it to your guinea pigs?" The sweet older woman behind the counter exclaimed, "Well, of course! They would just love that!"

Last year it became clear that our family was happiest when we had the most opportunities for animal encounters, so we up and moved from the city to rural New Jersey (yes, there is a rural part of New Jersey). Our backyard serves as a makeshift wildlife rehabilitation center, our home as a foster and permanent rescue for animals, and we're in the process of being able to provide sanctuary to rescued farm animals.

Our family's journey with animals is ongoing. I know mine is forever. Sure, my kids may lose some interest, but I'll be there to support them in any direction they care to go. I trust that they will always be inspired to help others.

By sharing what my family and I have learned and experienced over the years, I hope to inspire all animal lovers to learn a little more and do a little more. I've made mistakes with animal care with the best intentions. The more I learn, the better I do.

Our family's code is to live according to the golden rule: we do unto others as we would have them do unto us. I'm proud that I've been able to show my kids that taking care of others has value and is profoundly empowering and necessary. I feel whole, and I'm happy to be able to show that to my kids as well.

ANIMALS AT
Home

I T WAS 1995 AND I WAS TWENTY-EIGHT and living in San Francisco, studying design with my then boyfriend, whom I will kindly refer to as "Jack Ash." To have flexibility for my studies, I had gotten a job filling candy machines in stores, gas stations, schools, and any number of other places along a route from Santa Cruz to Oakland. The job allowed me to set my own hours and came with my very own van filled with candy!

While driving through various towns on my candy route, I found myself unable to pass any animal shelter without going inside. I felt compelled to share a moment with every one of the dogs caged there. I would try to look into each dog's eyes, and as I did so, my heart would break into tiny pieces. (Luckily, I'm allergic to cats, or my trips would have taken even longer.)

This candy gig was supposed to be allowing me more time for my studies, but my growing obsession was beginning to be counterproductive to the plan. Why did I feel so unable to pass an animal shelter without going inside? And why did my visits only leave me feeling sad and defeated?

Life with Jack Ash had driven me to seek therapy, so I put these questions to my therapist. Her belief was that I profoundly identified with these shelter dogs because when I was young, my parents had a very tumultuous marriage and I believed that if I could just "be good" and "do better," my parents would stop fighting. If I cleaned my plate, put my toys away, and went to bed without arguing, there would be nothing for them to scream about. But no matter how good I was, I believed it was never seen. These shelter dogs, through no fault of their own, had ended up in a very bad situation. I desperately wanted—needed—them to know that someone saw their goodness.

Once I was able to get a handle on the "why" of my visits, I was able to control myself more and restrict my sad ritual to a monthly trip to my local shelter. My apartment building didn't allow dogs, which was the only thing keeping me from being the doggy Mia Farrow I am today. But one fateful day in May at the SFSPCA, my eyes met with a dog's eyes that rocked me to my core. I found my doggy soul mate in a chocolate-brown gaze. I was going to convince my landlord to allow a dog—specifically, this wiry white mutt with brown spots named Enzo.

Fortunately, my landlord lived on my street and every day could be seen taking his two small children for a walk. I shamelessly plotted to ambush him in the company of his family. Surely his children had been begging for a puppy of their own—they were children, weren't they? Thanks to me, that itch would be scratched, and he wouldn't have to deal with housebreaking, chewing, barking, or, well, anything! All he needed to do was say yes, and his children would be invited to visit whenever they liked.

He gave his blessing.

The next day, Enzo came home to live with Jack Ash and me. All was finally right with the world—except that damn Jack guy was still around! When Jack explained to me that the best I could ever be was pretty, not beautiful, because there was too much space between the top of my lip and the bottom of my nose . . . I stayed. When he constantly corrected my grammar . . . I stayed. When he felt entitled to be highly critical of others based on the impressive number of books he had read, while ignoring his own lack of emotional intelligence . . . I stayed. But man oh man, when he accidentally left the front door open and Enzo ran into the street, narrowly escaping being hit by a car . . . it was time for me to go.

A week later, Enzo and I were on a plane heading to New York. Jack Ash admitted that he knew the minute I adopted Enzo it would only be a matter of time before I left him. In Enzo I had found a companion who loved me and made me feel beautiful despite that extra space above my lip. Enzo loved going for brisk walks as much as he loved lying in bed all day while I read. He was ecstatic to see me every time I came home. He adored my cooking, and he would turn himself into a fur-covered hot water heater when it was time to go to sleep. As long as we were together, he was happy. I saw the good in Enzo, and Enzo saw it in me. Things had gotten better for both of us, and thanks to Enzo, I started to trust and understand that life could be good. Really good!

When we hear the phrase "rescue animal," we tend to think of a dog or cat being rescued by a human. But when Enzo came into my life, I learned that more often than not, the rescued animal is the human, and the rescuer usually has four legs (or sometimes three).

Giving Back to My Furry Family Members

WHEN I WAS young, I believed in magic. I believed that if I wished hard enough, one day my dog, Muffin, would speak to me. I'd go to bed staring at him . . . waiting . . . waiting . . . waiting. I was obsessed with my dog and read every dog-related book I could get my hands on. Eventually I realized that my dog had been speaking to me all along. I just didn't yet know "dog-ese."

Over the years we both became proficient in speaking each other's language. Every morning at breakfast as Muffin nudged his way onto my chair to lie pressed against my back, he was saying, "I love starting my day with you, and I'll miss you while you're gone." If I sat too long doing homework, he'd come over, bow his head, and raise his butt high in the air to declare, "You've worked hard enough. It's time to play!" He'd lick my face to let me know that he, too, thought strawberry ice cream was delicious. At night in bed, he'd lay his head on my chest and whisper, "I love you very much. You are my best friend." I'd stroke his back and tell him I felt the same.

THE DOG:

I Have to Admit,
Hands Down My Favorite

I NEVER TAKE MY relationship with my dogs for granted. What if the canine species hadn't fallen so neatly and fantastically into domestication? What would I have done without my four-footed soother, my crutch, my inspiration, my motivation, my confidant, my best friend?

Sharing a home with a dog has been proven to have great emotional and physical benefits. Unfortunately, I can't report that I've become more physically active because of my dogs, but as I approach my fifties, I have discovered that they do have a certain effect on my looks. People often comment on my youthful appearance. I let them assume that I slather on La Mer moisturizer every night and go in for frequent Botox injections. I can only reveal to certain people (the types of people who'd be reading this book) that I actually attribute my youthful glow to liberal applications of dog saliva. Dog kisses get the oxygen in my skin moving and leave behind a certain glisten. Next time you consider buying a super-expensive moisturizer, stop and consider adopting a dog instead. Even if it doesn't give your life new meaning, at least you'll have dewy fresh skin.

As for the emotional impact of animals on my life, this entire book speaks to that. From a very early age, I recognized that animals gave sanctuary. Right from the start, my childhood dog, Muffin, comforted me during the night when so many fears would arise, whether it was the ghosts in my closet or the thought of facing Joannie Keaggy the morning after she'd challenged me to a fight. I held on tight to my little furball and fell asleep, only to wake up and realize that the ghosts had never appeared, and Joannie had changed her mind and decided she wanted to be best friends instead.

If guardian angels really exist, mine don't have wings. They have wagging tails, soft pink bellies, and terrible breath.

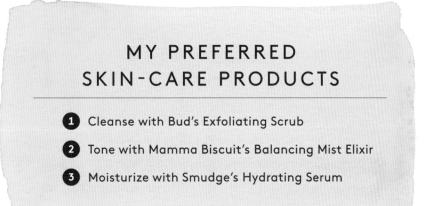

MY PREFERRED SKIN-CARE PRODUCTS

1. Cleanse with Bud's Exfoliating Scrub
2. Tone with Mamma Biscuit's Balancing Mist Elixir
3. Moisturize with Smudge's Hydrating Serum

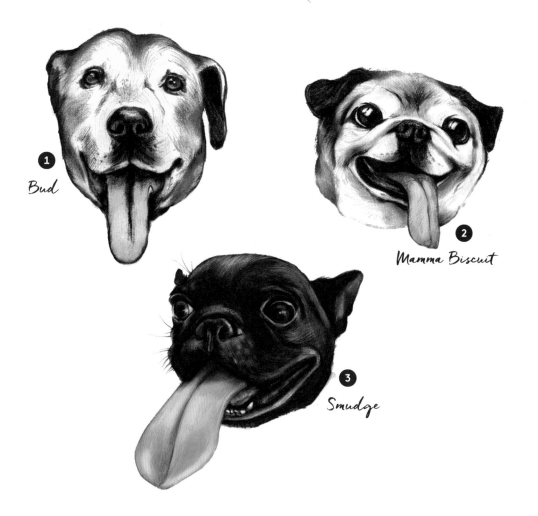

1 Bud

2 Mamma Biscuit

3 Smudge

DOG-ESE:

Learn to Speak a Dog's Language, as Modeled by My Foster Dog Mr. Fantastic

ONE OF THE greatest gifts you can give to any animal is to respect the species. Dogs and humans have been cohabiting for many years now, and dogs deserve most of the credit for the love affair. They have a long history of being extremely adaptable friends, open to learning all they can from us, and working hard to fit into our world. Every dog is an individual, with a very particular personality. I've learned an incredible amount from every dog I've adopted, as well as from those I've fostered. Without the mutual effort to understand one another, much can get lost in translation between human-ese and dog-ese. We owe it to our canine friends to learn a little of their language.

LISTENING

A subtly tilted head and a softly wagging tail say, "I'm all ears!"

NERVOUS

A dog crouching low to the ground with his tail between his legs is begging, "Get me out of here!"

AGGRESSIVE

Forward ears, exposed teeth, and
a tense body are saying, "Step back!
I don't feel safe."

FRIENDLY

There's the pointed-ears
and wiggly-body "I'm
feeling happy" look!

PLAYFUL

We all know the "Really, you
should drop everything and throw
me that ball" look, known as the
play bow position.

ALERT

Dogs have an incredible sense of hearing.
An erect neck and perked-up ears tell us
that they are well aware of what's going on.
Amazingly, dogs can locate the source of
a sound in six-hundredths of a second.

ADVANCED DOG-ESE:

Calming Signals, as Modeled by My Foster Dog Ms. Georgia Peach

WHALE EYE

A hard, focused stare that shows a lot of the white part of a dog's eyes pretty much means, "Give me space, *now!*"

DOGS HAVE AN expansive body language vocabulary that often goes unnoticed by the untrained eye. Like many people, most dogs hate confrontation, and they have myriad ways of communicating this to other dogs. These behaviors are often referred to as "calming signals" because they are a dog's attempt to calm another dog or himself. The dog is also attempting to communicate that he would like to prevent or resolve any potential conflicts. If your dog responds to the approach of a person or of another dog by averting his gaze, yawning, or scratching, these behaviors may indicate social anxiety (especially if they are clustered together). When your dog is stressed, you want to create distance from what is causing his stress, reassure him (in a happy, positive tone), and offer a valuable food treat (one your dog likes a lot). Remember that stress-causing event, and do your best in the future to avoid it, or be ready to train your dog to become more comfortable in that situation. If you feel your dog has a high degree of generalized fear or anxiety, seek the advice of a board-certified veterinary behaviorist.

AVERTED GAZE

Turning away and looking
in another direction is the
doggy equivalent of saying,
"I'm trying to remove myself
from the situation without
making a run for it."

LIP LICKING

There's the "finished my
last treat" lip lick, but
there's also the short,
repetitive lip lick when
no food is involved. This
can indicate that your
dog is feeling stressed or
uncomfortable.

RELAXED

Soft ears, soft tail, and softly
blinking eyes signal all-around
soft, yummy goodness.

SNIFFING THE GROUND

Who hasn't met a dog that doesn't love to put her nose
to the ground to get a good whiff of all the fascinating
smells found in every small patch of earth? But sometimes
this otherwise pleasurable behavior can be a dog's way of
saying, "I'm busy sniffing, and I've got no time for trouble."
Think of it as the dog-ese version of sticking one's head in
the sand: "Nope, nope, nothin' goin' on here."

Note: This is a general guide to dog
body language. Every dog has a unique
range and style of expressiveness via
normal canine body cues. A dog's
"vocabulary" can be influenced and
even impeded by human interventions,
such as breeding for particular body
structures and through procedures like
ear cropping and tail docking.

YAWNING

A yawn can simply mean "I'm
tired," but dogs sometimes
yawn in an attempt to
inhale more air to calm
themselves in a stressful or
confrontational situation.

EARS AND TAILS:

The More Appendages, the Better

IT WAS THE week before Thanksgiving and my kids and I were volunteering at Animal Haven shelter in New York City when a precious puppy named Christopher Robin was brought in. Days earlier, someone had cut off Christopher Robin's ears with scissors and left him for dead in a garbage can. It was surmised that whoever had done this probably hadn't been happy with the results and considered the dog dispensable. I'd have to use too many curse words to describe what I think of this, so I will just go on from here.

Christopher Robin was safe at the shelter, still very frail and frightened, when it was decided that a quiet foster home rather than a busy shelter was probably a better place for the initial phases of his recovery. This pup was going to need round-the-clock hugs, and luckily, my family of huggers was an approved animal foster family. We brought Christopher Robin to my house, which was about to be full of tenderhearted suckers arriving for a holiday feast. It was there that I'd launch my diabolical plan to keep Christopher Robin in my extended family. I knew that my sweet friends and family would be fighting over who was going to give Christopher Robin his forever home—and indeed that is exactly what happened!

Christopher Robin suffered horribly from the barbaric removal of his ears, and throughout his life there will be times where he will be unfairly judged because of how he looks. People may assume that he is an aggressive fighter, as dogfighters cut off (crop)

My nephew Christopher Robin

dogs' ears to make them less vulnerable in a fight. Other dogs may view stiff, immobile ear nubs as an indication of impending aggression, so these dogs can be at greater risk of being attacked.

It's hard to imagine the kind of cruelty Christopher suffered at the hands of a scissors-wielding human. Sadly, though, formal surgeries to crop dogs' ears and dock their tails (removing the hair from the tail before clamping and cutting it down to the desired length) happen every day in veterinary clinics. There has been no proven science to support the argument that these procedures prevent injury and infection. The decision to remove these vital appendages comes down to a fashion choice. Just as there's a purpose for every appendage on the human body, so it is for our dogs. Sometimes I even feel as though I could use more appendages, so the idea of removing any seems thoroughly misguided!

WHAT CAN WE DO?

You can start in your own home by making a promise not to put your furry family member through that pain. Discuss the issue with your family and friends. If your local veterinarian still performs this procedure, write a letter to him or her and explain why you'd like this practice not to happen in your community. Ask for the change you want to see. Write to local senators, representatives, and assembly members who have influence over the specific bills related to these issues.

The Animal Legal Defense Fund is just one organization working to fight these issues and offering resources for getting involved and making a difference.

THE CAT:
Our Relationship Was Complicated

My original competition:
Stanley

IN THE FALL of 1995, I met the man of my dreams. The only problem was, he had a cat. To protect this cat's identity, let's call him Stanley. Now, I want to be clear: I love cats, but my allergies do not, and I was confident that bouts of hives and phlegm were no way to start a relationship. So I very carefully avoided spending any extended amount of time in my potential husband's apartment, hoping that my weird behavior would be seen as mysterious and virginal.

But of course, the moment arrived for an overnight playdate. A night of romance and antihistamines. All was going beautifully, love was blooming, and sneezing was at a minimum. At some point after dinner, my new man had to take a business call, and I retreated to the bathroom for some freshening up and a hit of albuterol. That's when I saw out of the corner of my eye... Stanley. Hopped up on asthma meds and feeling cocky, I reached down to coo sweet nothings into the little kitty's ear—whereupon he leaped up and attached himself to my face with his claws.

Not wanting to disturb my man in the middle of what I imagined could be a career-altering phone call, I silently attempted to remove Stanley's death grip on my face. But no sooner would I detach one paw from my forehead than another would attach itself to my chin. Stanley was duct tape with fur, claws, and a vendetta. By the time the call ended I was bloodied, covered in hives, and hiding in the bathroom with a cold compress to my face. Now I was full-on mysterious.

I quickly figured out how to handle the hives. And there was a side benefit: the constant dosing of antihistamines made me seem relaxed and comfortable with myself. The relationship progressed; my man was falling for me hard. In the evening he'd head out to tell jokes, and I'd get to work snooping, as I was convinced he was going to leave me for an allergy-free supermodel. Stanley was always there, watching me, judging me for my pathetic and dishonest behavior.

Eventually I relaxed in my relationship with my future husband, and Stanley and I grew to love each other deeply. Stanley promised never to out me about my privacy indiscretions. We spent countless hours cuddled up on the sofa getting to know each other. I even welcomed Stanley to sleep on my pillow. The hives were all worth it.

Often people assume that if they know what their dog likes, then they know what *your* cat likes. *Au contraire*, my friend! I made that mistake with Stanley.

"Those are his personal magazines. Even I leave them alone."

"The girlfriend who was here before you wasn't allergic."

"You clearly are a dog person. I can tell by how you rub me."

CAT-ESE:

Learn to Speak a Cat's Language, as Modeled by Lisel's Cat, Min

C ATS, LIKE DOGS, use their ears, tails, mouths, and bodies to give us hints about how they're feeling, but they don't always use them in the same way. A dog wagging her tail loosely at medium height might indicate friendliness, but a cat moving her tail back and forth in a twitching motion might indicate that an unfriendly encounter is imminent. We all know what happens when we assume. Let's not sell our kitty friends short; they have so much to say. Let's listen.

HAPPY

No, you're not imagining things. That really is a smile. This kitty is happy to see you and if she's slowly blinking at you, she's giving you a "kitty kiss." *Purrrrr!*

Note: This is a general guide to cat body language. Every cat has a unique range and style of expressiveness using normal feline body cues.

RELAXED

A relaxed cat who approaches you and exposes her belly without hesitation is showing a sign of trust and calm.

ANGRY

A firmly pointed tail and enlarged pupils should be a clear sign to stay away. And if that's not enough, a serious hiss says it all.

HUNTING MODE

Hunting is one of a cat's most natural behaviors. Cats on the hunt keep their body low and stalk their prey in silence and with determination. Steady . . . steady . . . pounce!

FRIGHTENED

A kitty with a rounded back and fur standing on end is saying, "Protect me. I'm scared!"

NERVOUS

The typical "scaredy-cat" keeps her tail between her legs, crouches low, and is bright-eyed and fluffy-tailed.

CAT CLAWS:
Nature Intended for Them to Be There

P UDDIN' WAS SURRENDERED to Animal Haven shelter for scratching her original guardian's furniture. Due to my allergies, we weren't able to adopt her, but my daughter made it her mission to find her a new home. She set Puddin' up for success by supplying her with opportunities for letting her claws loose. She bought a simple cardboard scratch pad and decorated it with duct tape to personalize it just for Puddin'. It may be hard to understand, but cats really aren't trying to drive you crazy by scratching your belongings. Their claws are an essential part of their defense system, and they need to scratch surfaces to loosen the sheaths on their claws and allow new ones to emerge, as well as to get rid of dead skin. Providing cats with places and spaces where they can scratch to groom their claws, along with regular nail clipping, helps cats to stay healthy by preventing claws from growing into their toe pads, which could lead to infection. Scratching is also one way that cats mark their territory with their scent, courtesy of glands in their toe pads. For all of these reasons, scratching is a natural habit that requires an appropriate outlet.

My daughter's obsession, Puddin'

All too often, however, cat owners decide that declawing their cat is the solution to stopping "destructive" scratching behavior. Because these procedures are performed so routinely, they can often be trivialized. As a result, many loving and caring people are unaware of just how painful and debilitating declawing a cat really is.

In humans, fingernails grow from the skin. In cats, claws grow from the bone. When a cat is declawed, those bones are essentially amputated so that the claws cannot regrow. The tendons, nerves, and ligaments that allow normal function and movement of the paw are severed and the cat is left crippled and robbed of this primary means of defense. Declawed cats can suffer permanent lameness, pain, and arthritis, as well as behavioral problems stemming from these effects. They may avoid using their litter box due to the pain of digging in the litter, and they may resort to biting because their first line of defense is no longer an option.

Declawing is never the right answer for handling your cat's long claws, but this is not to say that kitties don't need a regular manicure and pedicure. Getting unsnagged from a knit blanket—or from your favorite sweater—is no fun for anyone.

WHAT CAN WE DO?

As an alternative to declawing, house cats can be redirected from furniture to more desirable objects, like a scratching post or mat. These objects will satisfy a cat's desire to use her claws and will help to file the claws down on a regular basis. Another great product is Soft Paws, little rubber nail caps that can easily be slipped onto your kitty's claws. Soft Paws requires that the tip of the cat's nail be trimmed just slightly to allow the cap to fit down to the base of the nail. Using a nontoxic adhesive, the nail caps are secured onto the paws, lasting 4 to 6 weeks and allowing kitty to scratch, stretch, and partake in all other natural behaviors without any damage.

PRACTICE THE ART OF ANIMAL MASSAGE

T HERE'S LITTLE I love more than feeling that I have connected with an animal. Touch has always been my most effective tool for expressing my affection and showing my respect for my animal friends. Before meeting a new species, I research how that particular creature prefers to be approached and, unless the animal is to be left in the wild, touched. I've found that my attention to detail is highly appreciated.

Let's face it, when done right, there are few things more enjoyable than a good massage, and when it comes to animals, massage can be very therapeutic for animal and human alike. Of course, not all dogs and cats like to be touched, and massage is a very particular kind of touch. So before you dive in, make sure you are aware of the individual animal's likes and dislikes with regard to touch.

ESSENTIAL DO'S AND DON'TS
FOR MASSAGE OR TOUCH TIME

DO	DON'T
• Make sure you are relaxed before massaging your cat or dog.	• Approach your cat or dog from above the head or directly in the animal's face.
• Choose a location that has few distractions for both of you.	• Use force or press on the animal's stomach.
• Allow time for your cat or dog to sniff and lick your hands and to prepare for your touch.	• Pull on whiskers, ears, or tails.
• Use a soft tone of voice to praise and talk to your cat or dog during massage.	• Attempt to massage when the animal is eating or otherwise occupied.
• Pay attention to your furry friend's body language during touch time. If you see signs of tension, slow down or stop.	• Massage an aggressive animal or an animal whose body language and personality you don't know well.
• Watch for signs of stress, such as yawning, lip licking, and averting the face, or the more obvious indicators of needing more space, such as trying to move away.	• Ever massage an animal you do not know, and even with your own furry friend, be sure to act carefully and considerately.

NOT SO RUFF:

Six Ways to Massage Your Dog, as Modeled by My Dog Barkly

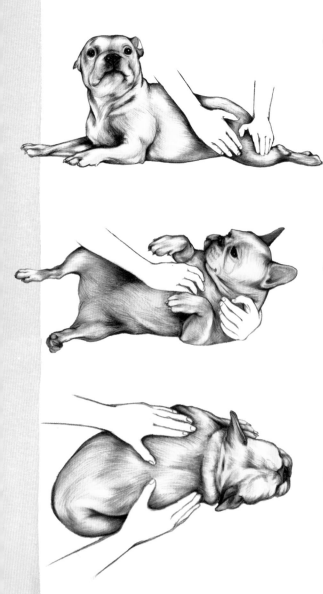

1 Thumb-rub the thighs: Extend your thumb away from your fingers and place your hand so that your thumb is in the middle of the dog's thigh and your fingers point toward the outside of the leg. Starting high near the body and using mild pressure, slowly rub small circles with your thumb, gradually moving your hand down the leg, ending near the paws.

2 Stroke the chest: Bring your hands to your dog's chest and wait for him to lie down on his side or back. With your fingers in a loose claw shape, use your fingertips to lightly scratch along the chest and under the armpits in a circular pattern. If your dog enjoys this, try gradually adding pressure and speed.

3 Circle the shoulders: Gently placing both palms flat on either shoulder, use your finger pads to make small circles with mild pressure. Gradually make the circles bigger, changing directions and varying speeds.

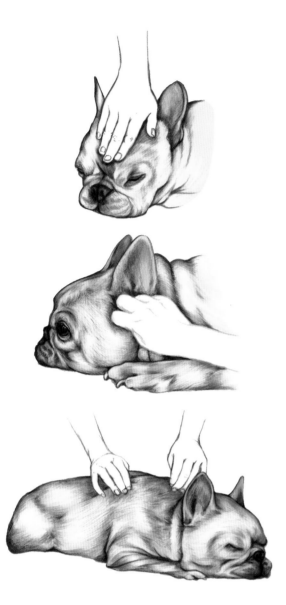

4 **Rub the forehead:** With your palm facing down and four fingers together and flat, rub your dog's forehead from the top of the head to the tip of the nose. Occasionally make small circles with your finger pads and rub up and down with light pressure.

5 **Nestle the base of the ear:** Ball your fist and use your knuckles to rub small, gentle circles at the base of your dog's ear, just below the opening. Keep your palm facing down and use light pressure. A dog's ears are sensitive, so pay attention to the pressure and move slowly.

6 **Gently tug at the scruff:** Grasp the loose skin on the scruff of your dog's neck or on other loose, furry areas of your dog's body and slowly squeeze your fingers and thumbs together while gently pulling upward. Try to alternate left and right hands, lifting the left up as the right lowers, and vice versa. Be careful not to pinch the skin at the end of the pull.

Note: Only give your dog a massage if he is relaxed and receptive. Soft ears, a slightly opened mouth, and a relaxed tail can be good indicators that things are going well. Avoid massaging a sick animal (unless recommended by your veterinarian), and do not massage an animal in place of veterinary care. Be sure to pay attention to the animal's body language when attempting massage. Animals will let you know whether they are enjoying it or not!

MEOW-SAGE ME:

Six Ways to Massage Your Cat, as Modeled by Animal Haven's Mykah

1 **Circle the sides:** Gently place both palms flat on either side of the cat's spine. Make large circles using your finger pads, pressing lightly. Alternate the move by simply using both hands to rub up and down the sides. Only add pressure on a downward stroke and not going against the fur.

2 **Pat the forehead:** Using four fingers together and flat, rub up and down your cat's forehead. Test different areas on the top of your cat's head, rubbing slowly with light pressure.

3 **Stroke the chin:** Lightly use your fingertips to make small circles under your cat's chin. Watch to see if your cat tilts her head back and points chin to sky in enjoyment. Then you can begin to move your fingertips gently up and down from the chin along the front of the neck, the jowls, lower jaw, and cheek area.

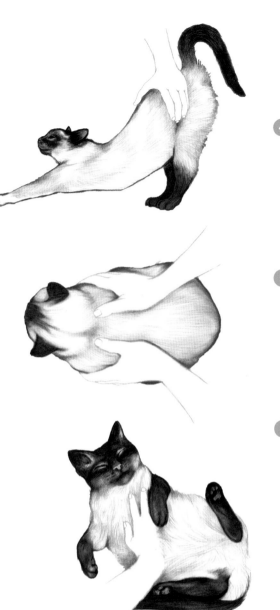

4 **Rub the rump:** Use four fingers to rub the lower part of your cat's back, applying mild pressure. Move back and forth horizontally with short strokes. If your cat raises her lower back, she is enjoying the stroke and the pressure.

5 **Massage the shoulders:** Place one hand on each of your cat's shoulders and notice the hollow area between and beneath the shoulder blades. Gently work around these areas, kneading in soft circular motions.

6 **Rub the chest:** When your cat has her belly exposed, rest your palm gently on her chest and move your fingers lightly around her pectoral muscles. Rub the front of her chest and around her shoulder blades.

Note: Only give your cat a massage if she is relaxed and receptive. Avoid massaging a sick animal (unless recommended by your veterinarian), and do not massage an animal in place of veterinary care. Be sure to pay attention to the animal's body language when attempting massage. Animals will let you know whether they are enjoying it or not!

MEET YOUR MUTT:

Let's Mix Up the Gene Pool, People!

WHEN I WAS pregnant with my children, I wanted to know their gender as soon as possible—not because I wanted one or the other, but because it felt comforting to eliminate at least one unknown. When folks are pregnant with the notion of getting a dog, I can understand why some might choose a specific breed in hopes of knowing exactly what they're in for. Choosing a specific breed may allow one to gauge the general look and size of the adult dog, and there are some behavior traits that hold true—herdin' dogs wanna herd; water dogs wanna swim! However, I have yet to parent two dogs of the same breed that weren't vastly different. "Pureness" of breed has never offered any real shortcuts in terms of knowing how to handle the dog. And, unfortunately, purebred dogs suffer a host of genetic defects and illnesses. Within any breed and even within every litter, there is a canyon-wide range of personalities. Just as every person is unique, so every dog is different.

Personality and compatibility between dog and guardian are the best predictors of a long-term love affair. It's not that different from humans, really. Clearly, one reason I chose my husband was for his outstanding good looks. But luckily for the long-term well-being of our relationship, he also has a calm nature, a bubbly personality, and a keen ability to recognize that I'm always right. Once you fall in love with a great personality, you're going to want to have that special individual around for a very long time.

Unlike purebreds, mutts come from an expansive and varied gene pool, which not only increases the likelihood of good health but also bodes well for finding just the right match for a long-term, committed relationship. The good news is that if you fall in love with the notion of a mutt, they are available for adoption right now at a shelter near you!

The next time you're at the dog run and others are boasting about their fancy purebred, consider stealing the limelight by boasting about your Fluffy-Tailed Shepherdshund.

The Everything Bagelhund

The White-Bibbed Snuggler

The Sweet-Faced Goodenboy

The Tiny-Headed Earhound

The Flap-Eared Gettongueinhund

The Sad-Eyed Shepherdmush

The Bearded Schnitzel

The Stump-Legged Beghound

The Bed-Headed Newyorkie

FIVE WAYS TO MAKE A DOG HAPPY

WHEN CARING FOR a dog, it's important to consider the various toys, treats, and accessories needed to meet his social, physical, and mental needs. Whether you're looking out for your own family pet or caring for a dog in a shelter, you can make a big difference in his quality of life by adding enrichment with your time, attention, and love.

① PROVIDE A VARIETY OF TOYS

Dogs enjoy having different types of toys to play with. Food-dispensing toys are good for mental stimulation, while chew toys provide an unlimited amount of entertainment and relaxation. My kids can spend hours designing and making toys when they know they'll get to see them enjoyed by appreciative shelter animals.

2 SET UP PLAYDATES

Dogs thrive when taken outside the home to experience new places and people. Organize playdates with other dogs. Visit an indoor or outdoor run for healthy physical and mental exercise. It's incredibly important that you make sure the run has safety rules in place and that all participants follow them.

3 PLAY HIDE-AND-SEEK WITH TREATS

Hiding treats around the house keeps a dog mentally and physically active. Dogs enjoy the search and appreciate the rewards they discover. Make your own home-made treats and sneak some greens into them to help the development of healthy blood and muscle tissue (see my Pumpkin Dog Biscuits, page 43).

4 KEEP TREATS HANDY

This makes rewarding dogs easy for the trainer as well as for any other person interacting with the dog. Make sure you can get your hand into and out of your treat holder quickly. Widemouthed jars are great for this reason, as are treat pouches that attach to your waist. The goal is for the dog to never know when reinforcement may happen, so you want to have the treats/toys readily accessible but not visually obvi-ous—otherwise the dog learns to focus on you when he sees the pouch and tune out when he doesn't. You also want to be able to mark the desired behavior with a treat as soon as the behavior happens. Be sure to label any jar or container of dog treats to stop heedless family members or unwitting visitors from digging in!

5 OFFER A VARIETY OF SENSORY ITEMS

There are various smells and sounds that can help dogs relax. Lavender and chamomile scents relieve stress, while classical music and recorded ocean waves can help manage behaviors associated with fear. Experiment with different sounds and smells to see what works for your dog in your environment.

A soft dog bed suited to your dog's size and age may also help to relieve stress and provide comfort. There are dog beds that offer orthopedic, heating, and cooling benefits. My high-energy and always-on-the-go French bulldog, Smudge, prefers a smooth, cool dog bed; my other Frenchie, Barkly, who suffers from frequent bouts of back pain, loves his heated bed; and my two pit bulls, Lil' Dipper and Scout, who love to sleep under the covers, truly appreciate their dog beds with cavelike roofs.

Craft

WATER-BOTTLE SOCK CAT

SUPPLIES

- Plastic water bottle (16.9-ounce)
- Long sweat sock

- Corrugated cardboard (enough to make two small ears)
- Needle and black thread

DIRECTIONS

1   Cover the plastic water bottle with the sweat sock so that the toe of the sock covers the bottom of the bottle.

2   Cut two cardboard triangles and push them inside the sock toward the toe to make pointed ears.

3   Sew the sock closed beneath the lower edge of the triangles to keep them in place.

4   Tie the top of the sock in a knot to hold the bottle in place.

5   Continue to tie knots down the remaining material until you reach the edge of the sock. These knots become the tail.

6   Use black thread to add feline facial features.

Bake

PUMPKIN DOG BISCUITS

I especially enjoy cooking with my kids when I know I won't have to eat what they've made. They love to bake pet treats, and I love knowing that the ingredients are natural and healthy. Wheatgrass and alfalfa powder provide dogs with energy while rejuvenating their blood, stimulating their circulation, and regenerating their liver. Mixing these powders with mint and parsley can also help to curb stinky doggy breath! Wheatgrass should be added to your dog's diet slowly—too much too soon can lead to vomiting and sluggishness due to its detoxifying nature. A 2-inch bone-shaped cookie cutter should yield approximately 18 treats.

INGREDIENTS

- **1 cup gluten-free flour (I use King Arthur)**
- **½ cup almond milk**
- **⅓ cup pumpkin puree**
- **2 tablespoons wheatgrass (or alfalfa powder)**

- **½ tablespoon baking powder**
- **1 tablespoon fresh mint, finely minced**
- **½ tablespoon dried parsley**

DIRECTIONS

1   Preheat the oven to 350°F and lightly grease a cookie sheet.

2   In a bowl, combine all the ingredients and mix until a smooth dough is formed. Turn the dough out onto a well-floured surface and roll it out to a ½-inch thickness.

3   Use a 2-inch cookie cutter to cut the dough into individual treats. Be sure that the treat is an appropriate size for your dog to prevent choking.

4   Place the biscuits on the prepared baking sheet and bake for 20 minutes, or until golden brown.

5   Allow to cool completely. Store in an airtight container in the refrigerator. The treats will be soft and chewy and should be kept in the fridge for no longer than 2 weeks.

Craft

T-SHIRT TUG TOY

SUPPLIES:

- T-shirt
- Tennis ball
- Scissors or a drill

DIRECTIONS:

1   Cut the T-shirt vertically into three 5-inch-wide strips.

2   Cut two holes in either side of the ball carefully, using scissors (or a drill).

3   Tie the ends of the T-shirt strips together in a knot.

4   Braid the strips into a rope, stopping halfway down.

5   Thread one of the loose strips of T-shirt fabric into one hole in the tennis ball and out the other.

6   Continue braiding the fabric strips until they are completely braided. The ball should be secured in the middle of the rope with a braid on either side.

7   Tie a knot to secure each end of the rope. Your dog should be able to take the ball in his mouth while you pull on either end of the rope.

Craft

WATERPROOF EASY-SEW DOG BED

SUPPLIES

- Oilcloth fabric (available at well-stocked fabric stores, or use an old, clean tablecloth)
- Scissors
- Straight pins

- Sewing machine
- Cotton thread
- Egg crate foam
- 1 king-size pillowcase

DIRECTIONS

1   Measure and cut the oilcloth into two rectangles, each approximately 20 by 36 inches, adding a ½-inch seam allowance to each side so that each rectangle is approximately 21 by 37 inches.

2   Pin together three sides of the fabric rectangles with the oilcloth sides facing each other, leaving one of the shorter edges open.

3   Sew the three sides together to make the pillow cover (see Note).

4   Turn the pillow cover right side out.

5   Cut the egg crate foam to size so that it will fit snugly inside the oilcloth pillow cover.

6   Insert the egg crate foam into the pillow cover. Depending on your dog's preference and the amount of room inside the pillow cover, you may choose to add one or more layers of egg crate foam.

7   Stitch the fourth side of the pillow cover closed.

8   Cover with the king-size pillowcase to finish your dog bed!

Note: For sewing oilcloth using a sewing machine, use a size-16 needle and regular cotton thread. Use a long stitch so as to not perforate the oilcloth too much, which will allow it to tear apart.

KEEP YOUR DOG SAFE:

Teach the "Touch" Cue,
as Modeled by My Dog Scout

TRAINING DOGS PROVIDES mental stimulation and makes it easier to keep them safe. The "touch" cue is an important skill for your dog to have, as is walking nicely on a leash, sitting and staying on request, and name recognition. These cues can help you prevent any dangerous or dicey situations, like wandering into oncoming traffic, approaching an unfriendly dog, or stealing your friend's sandwich.

I use the touch cue with my dogs all the time: when they spot a rabbit in the backyard, when I don't want them to overwhelm a guest at the front door or eat food that was accidentally dropped on the ground, or even to position them on the scale at the vet's office—whenever I need them to come to my side. It helps them to learn good behaviors and prevents unwanted ones. It's also a great way to have fun with your dog!

1. Prepare a bunch of small, pea-sized treats for your dog and put them in your treat pouch. Your dog will naturally be interested in you now!

2. Extend one hand to the side about six inches away from your dog's nose. Keep your fingers pointing outwards with your palm facing toward your dog. Look for your dog to orient towards your hand, and take a few steps away, encouraging him to approach you.

3 The instant your dog touches your palm with his nose, mark his good behavior by saying "Yes!" and offering a treat from your pouch. Think of the word "Yes" as a camera capturing the behavior and announcing to the dog, "Here comes the food reward!"

4 You may be asking: Why not just give the dog a treat without using a marker word? Dogs are instantaneous learners, and using a marker word is a more precise way to give timely feedback. Saying "Yes" is a promise that the treat is coming while avoiding the distraction of having visible food in your hand when you're training.

5 When your dog learns to approach once you present your hand, you are ready to attach a verbal cue. It's simple! Say the cue "touch" (just once), and then extend your hand to the side. Mark "Yes" and give him the treat when your dog touches your palm.

6 Slowly increase the distance between you and your dog while you continue to practice "touch." If your dog is distracted, he may need a little more motivation. Verbal prompting, such as saying "Whoop-whoop-whoop!" in an excited tone or making kissing sounds is a great way to gain your dog's attention and motivate him to succeed.

7 If your dog is still not responding, then you likely need more practice in less-distracting areas. Try not to give up easily. If you persist and give big rewards when he gets it right, he will respond faster and faster over time. You don't want your dog to think that responding to the cue is optional or acceptable to ignore. I've learned this the hard way with my children as well.

Note: "No motivation, no training" is the rule when it comes to training. However, food is not the only option for motivation. If you are not using food, it is important to reinforce the desired behavior in some way. (Opening the door to let the dog out, throwing a ball, neck/butt scratches, and praise are just a few other possible alternatives—depending on what motivates the individual animal.) Behaviors that are not reinforced will decrease.

POTTY TRAINING:

Surviving in a World of Unreachable Toilets, as Modeled by Animal Haven's Good Boy Larry

I HAVE HAD MANY dogs and have tried many methods to convince them not to pee or poop on my carpets. When success wasn't immediate, I was always the guilty party. Keeping to a consistent schedule and setting your pooch up for success are all on you, my friend, no matter how much you'd like to blame problems on the dog. A big reason animals are surrendered to shelters involves failures in potty training. Those failures are never the animal's fault but a function of inconsistent training or a medical condition. Let's just say the dog isn't ever the one who should be hit on the nose with a rolled-up newspaper. We can lower the number of animals in shelters by educating ourselves and giving our dogs the tools they need to live in our world of pricey rugs and human-designed toilets.

I implore you to use positive reward-based training. Rewards can include food treats, access to play, walks, door-opening services, and so on. Almost anything our dogs need or want on a daily basis can be used as a reward. I parent with tenderness, understanding, and consistency, not intimidation. Many approaches offered as "quick fixes," including shock collars, choke collars, or physical domination, have more in common with animal abuse than with teaching dogs to trust humans. Harsh, painful punishments disguised as training will cause behavioral fallouts, such as generalized fear, aggression, and anxiety.

Here are a few tips that have really helped me (and my carpets!) when it comes to potty training.

1. **Be positive.** Give your dog lots of opportunities to relieve himself in the desired place and give him loads of praise when he does. If your dog has an "accident," don't be a jerk and push his nose in it. Unlike humans, dogs have mastered the art of living in the moment and are instantaneous learners. If you find your dog's accident after the fact, it's too late. Don't bother addressing it; he won't know what you're yelling about.

2. **Keep to a schedule.** Develop a schedule that takes both your and your dog's needs into account. Evaluate it regularly, be flexible, and adjust it according to your puppy's activities, growth, and development. I currently rely on a great app called Pile and Puddle (pileandpuddle.com) to help me keep track of my puppy's business.

3. **Be considerate.** Young puppies (eight to ten weeks old) don't have much bladder or bowel control. In general, puppies are inclined to go when they wake from a nap, when they play or get excited, and shortly after they eat or drink. During the day, your active puppy may go every twenty minutes to an hour. Factor in these issues when considering your puppy's schedule.

4. **Remember that we all like to go before hitting the sack.** Give your puppy the chance to go before bed, so both of you rest comfortably through the night. A puppy is likely to need a potty break or two overnight, but some can make it through a six- to eight-hour sleep if they go a few hours before bedtime. Making sure your pup is tired and ready to sleep when you go to bed is helpful as well.

5. **Know that things will get easier.** As dogs get older, they can hold it for longer periods of time. Families that work all day should have a dog walker or a family member come home twice to give their dog a walk and a potty break. While puppies generally poop three to four times a day, most adult dogs only poop once or twice a day—which could be a good reason to consider adopting an older dog.

FIVE WAYS TO
MAKE A CAT HAPPY

INDOOR CATS DEFINITELY fare better than free-ranging feral cats. In fact, indoor cats live on average from fifteen to eighteen years, while outdoor cats typically live no more than three. But let's face it, indoor life can get pretty boring for our feline friends. Keeping an indoor cat happy means catering to all the fun she misses outdoors and providing the best opportunities for behavioral outlets.

1 EXCITE THE SENSES WHILE GETTING COZY

Cats are sensory-driven and require stimulation of all five senses. Find a way to incorporate cats' desires to climb and perch high, smell different scents, taste a variety of flavors, watch and listen to outdoor activity, and feel different materials under their sensitive paws. Balance this fun and excitement by providing cozy snuggle spots.

2 PLANT CAT-FRIENDLY GREENS TO EAT

Cats are inclined to investigate and digest plants, but houseplants can be filled with fertilizers and chemicals that can disrupt cats' stomachs, and some plants are poisonous to cats. Growing cat grass and cat mint for them to eat instead is a simple solution for keeping your other houseplants cat-free, satisfying cat cravings, and making healthy additions to your cat's diet.

3 PROVIDE SAFE HUNTING OPPORTUNITIES

Hunting is a natural behavior for all cats, indoor and out, as they are hardwired to use their physical and mental abilities to catch prey. A cat teaser (such as a toy on a string) keeps cats active, gives their muscles a workout, helps their bones to stay strong, and allows them to develop a healthy appetite.

4 SUPPLY A VARIETY OF TOYS AND SCRATCH PADS

Batting around a ball or other small toy provides activity and fun for cats. Playing with different toys helps them to practice their coordination, adjust to the speed of moving objects, and gauge distance by pouncing. Shiny materials and cardboard items are also hits with cats. However, it's important to never leave a cat alone or unattended with toys that contain string, yarn, ribbon, or other similar items that can easily be swallowed, as they can lead to dangerous situations, including becoming tangled in a cat's intestines.

5 HIDE FOOD AND TREATS

Indoor cats will appreciate the mental challenge of "hunting" for their meal. Hide food in different areas of the house and allow them to sniff and search for their meals and special treats.

Grow

CAT GRASS

SUPPLIES

- Small potting container
- Organic potting soil
- Cat grass seeds
- Water
- Plastic sheet (Saran Wrap)

DIRECTIONS

1   Fill the container with organic potting soil, stopping 1 inch below the top edge.

2   Spread a handful of cat grass seeds on top and cover with a layer of potting soil.

3   Water the seeds by pouring ½ cup of water into the pot.

4   Cover the potting container loosely with the plastic sheet, ensuring that the covering is not so tight that some air cannot reach the soil.

5   Hide your pot in a dark, warm location until the cat grass begins to sprout. Once the grass is about 2 inches tall, remove the plastic sheet and relocate your pot to a sunny spot.

6   Let them eat grass! By the time the cat grass is 4 to 5 inches tall, your cat will want to dig in. It may take 7 to 9 days for the cat grass to grow to the perfect length before your animal is ready to approach. Don't fret; it'll happen!

Craft

HANGING BIRD CAT TEASER

SUPPLIES

- Printer
- Canvas fabric, 8½ by 11 inches
- Scissors
- Decorative fabric, 8½ by 11 inches
- Straight pins
- Needle and thread
- Polyester stuffing
- String
- Wooden stick, 4 to 5 feet long

DIRECTIONS

1   Find three bird images and print them onto the sheet of canvas.

2   Cut out the birds, leaving a 1-inch border around each image.

3   Pin the canvas birds to the decorative fabric.

4   Cut the decorative fabric in the same shape and size as the canvas birds.

5   Match each canvas bird with each decorative fabric bird, so you now have three matched pairs.

6   Pin each pair together, with the decorative fabric and the bird image facing inward.

7   Sew each pair together around the edges, leaving an opening large enough to fit two fingers for adding stuffing.

8   Turn the material right side out so that the image of the bird is visible.

9   Fill each bird shape with stuffing and sew the open edge closed.

10 Poke a hole near the middle of each bird toy using your needle, and thread string through the hole.

11 Hang the bird pillows on the stick by knotting the strings around the stick.

Craft

CARDBOARD PLAY PALS

SUPPLIES

- **Empty toilet paper or paper towel rolls**

- **Scissors**
- **Colored duct tape**

DIRECTIONS

1   Collect empty paper rolls of different sizes.

2   Cut patterns along the rolls, spiraling from top to bottom or adding frills on either edge.

3   Stick colored duct tape along the rolls as decoration.

Craft

CATNIP SEA CREATURES

SUPPLIES

- Black fabric marker
- Sheet of colored felt, 9 by 12 inches (1 sheet per creature)
- Scissors
- Straight pins
- Needle and thread
- Polyester stuffing
- 1 teaspoon dried catnip (per felt toy)
- Rickrack trim
- Pinking shears
- Sheet of white felt (for eyes)
- Fabric glue or hot-glue gun and hot glue sticks

DIRECTIONS

1   Use a fabric marker to draw the shape of your sea creature on felt. Be sure to allocate space for a ¼-inch border.

2   Cut out the shape and pin it to another piece of felt, then cut around the shape once more. You should now have two identical felt shapes.

3   Pin the two felt pieces together, leaving a 2-inch opening at the bottom for stuffing.

4   Sew together with a needle and thread, leaving a ¼-inch border around the sides.

5   Fill with stuffing and catnip.

6   If making an octopus, insert pieces of rickrack ½ inch inside the opening. Sew opening closed securely to stop the catnip and rickrack from falling out.

7   If making a fish, sew the opening closed and use pinking shears to trim the exposed seam allowance.

8   Cut small circles out of white felt to use as eyes. Glue in place using fabric glue and draw pupils using a black fabric marker.

SIGNS THAT YOUR CAT IS CUCKOO FOR CATNIP

CATNIP CAN RANK pretty high on the list of feline treasured items. Catnip stimulates your cat's senses because most cats are highly susceptible to the oil in the stems and leaves of the catnip plant. Whether it's in the air, in cat toys, in the garden, or in your cat's food, the effect of this simple herb on most cats is intense.

Catnip is part of the mint family and can easily be grown in your yard or in your home and dried to be stuffed inside toys or spread over various cat-friendly surfaces. Of course, all cats are unique and will react to catnip in different ways. I am one of those people who doesn't enjoy feeling out of control and stupid, and not every cat loves going cuckoo for catnip. Some cats simply couldn't care less about catnip. Catnip sensitivity is passed from generation to generation. In fact, an estimated 50 percent of cats are unresponsive to the wonders of catnip. Pay close attention to what your cat's body language is saying. Hopefully your kitty will be digging it!

HOW TO DRY CATNIP

1   Cut the plant at the base using scissors. Always cut more than you think you will use, as the plant will shrink through the drying process. Five inches is the shortest stalk you should use.

2   Bunch the stalks together and tie a piece of yarn tightly around them.

3   Hang the bunch upside down in a dark, dry area. Be sure to limit the amount of light exposure, as the plant will lose its potency if exposed to light.

4   Wait 4 to 8 weeks, or until the catnip is completely dry. Cut it up, put it in homemade cat toys or scatter it on your cat's scratch pad, and watch your kitty go crazy!

1 "BACK OFF!"

On occasion, you may be faced with a cat who is having a highly stimulated response to catnip—an aggressive, swatting, clawing creature that is not reacting well to the smell or the taste of the plant. Luckily, this episode will last only a few minutes.

2 "WOW!"

Running, rolling, flipping, and rubbing are signs that your cat has definitely taken a sniff of catnip's aromatic essential oil. The smell of catnip is said to release "happy" pheromones that travel straight up to the brain and make cats crazy!

3 "WHOA, DUDE."

If your cat is zoned out and off in a world of her own, she's most likely taken one too many bites of the catnip plant. Eating catnip can cause a cat to mellow out and is often prescribed to stressed cats who need help to calm down.

Animal Haven's "mini lion," Ludwig

Providing a
Forever Home

THERE'S NOTHING LIKE bringing a new animal into your home. The joy and excitement can make us wish we could take more home and, when we adopt from a shelter, save them all. But the truth is, we can save more animals and give them homes just by choosing shelters over pet stores. When we adopt a companion animal, we're not only saving that individual, but we're also making room in the shelter for another to be saved.

While shelters are often strained by limited budgets and by the sheer number of animals they must take in, they do their best to provide an optimal temporary environment. They care for every breed, every kind of wonderful combination of mutt, every size, every shape, and every personality. If you're looking for a hypoallergenic pet, they've got it. Looking for small and quiet? They've got it. Looking for energetic and loving? They've got it. When you adopt from a shelter, not only are you providing a loving home to a fabulous animal in need, but you're also supporting the people who genuinely care about the welfare of your pet.

GIMME SHELTER:

Adopt, Don't Shop

MANY PEOPLE PURCHASE animals from pet stores on impulse or because they assume, incorrectly, that shelter animals are "damaged goods." This couldn't be further from the truth. Certainly, some shelter animals have not had the benefit of a loving home and education (that's what you're there for!), but animals end up at shelters for all sorts of reasons. Some are the victims of divorce, illness, allergies, a new baby, inexperienced owners, a move that didn't include them, or the death of their caregiver, among many other circumstances. Regardless of why they ended up in the shelter, most rescue animals are loving beings, grateful to have a second chance at a happy life in a forever home.

My days working as a veterinary technician showed me that purebred dogs from puppy stores were just as likely, if not more so, to have behavioral problems as mutts from the shelter. And shelters are equal opportunity homes for purebreds, mutts, pups and kitties, and seniors. In my experience, an animal's presence in a shelter often says a lot more about the person who surrendered the animal than about the animal. We all know that not all people are winners! The most important thing that all shelter animals have in common is that every animal is waiting for a responsible and loving rescuer.

Shelter animals with the most daunting behavioral issues, such as extreme fear or aggression, are usually euthanized, especially if there is a bite history. Sadly, however, animals with absolutely no serious behavioral problems are euthanized as well, due to lack of shelter space and resources.

So the next time you're looking to add another member to your brood, check out Petfinder (www.petfinder.com), visit your local shelter, and research rescue groups, and I'm positive you will like what you find!

WHAT EXACTLY IS A PUPPY MILL?

Your Local Pet Store Won't Tell You

PUPPY MILLS ARE large-scale commercial dog-breeding operations that house dogs in overcrowded and unsanitary conditions. Puppy mill workers are more concerned about the profit to be made as a result of breeding and selling dogs than about the dogs' well-being. More often than not, the dogs are kept in wire cages and can spend their entire lives crammed indoors with inadequate veterinary care, food, water, grooming, sun, or socialization. Female dogs are bred at every opportunity without much time between litters and are put down once they can no longer produce puppies. Puppies are often riddled with hereditary defects because of an operator's failure to remove sick dogs from breeding pools and because of inbreeding.

As the public has become more educated about puppy mills, pet stores have become more savvy in disguising where their puppies come from. Mills have created elaborate online presences that trick people into thinking they are dealing with responsible breeders. If people adopted from shelters instead of buying from pet stores or online breeders, millions of dogs would be saved from being euthanized each year, not to mention the thousands of adult dogs that would be spared from puppy mill life.

Many folks are still unaware of the terrible conditions for parent dogs that are left behind in puppy mills once their pups are shipped away. They are also unaware of the dangerous conditions that puppies endure after being taken from their mothers, many not surviving the move and others suffering social, behavioral, and health issues. The ASPCA believes that if we stop buying from pet stores, and they become forced to sell puppies at reduced prices, then eventually they will keep fewer puppies for sale. Ultimately, this would translate to mills' producing fewer dogs, and hopefully one day becoming nonexistent.

SPAY AND NEUTER:
Our Biggest Hope

The Humane Society of the United States estimates that animal shelters care for 6 to 8 million dogs and cats every year in the United States, of which 3 million to 4 million are euthanized. This is a staggering number. We have got to do something. There are too many animals needing homes and not enough homes to provide adequate care, shelter, and love. Spaying and neutering are our best hope for beginning to pull these numbers down and give more shelter animals a chance at life with a family. Here are some of the additional benefits of spaying or neutering a cat or dog.

- They can reduce the animal's desire to urine-mark.

- These procedures have been proven to reduce an animal's risk of uterine cancer, testicular cancer, prostate cancer, and other cancers of the reproductive system.

- They can decrease dogs' and cats' desire to roam. Roaming can put animals in jeopardy of becoming lost and ending up in a shelter, being hit by a car, getting into fights with other animals, or falling into the hands of people who will abuse them or use them for profit through breeding or dogfighting.

PIT BULLS:

Shelter Animals with a Bad Rap

C LEARLY, CERTAIN DOGS are more physically powerful than others and can be visually intimidating to some people. If you choose to share your life with a larger dog, in addition to the extra-large portion of love you'll get, you need to be prepared to accept an extra-large portion of responsibility. People may be tolerant of a little dog that is a lot out of control but often have no tolerance for a larger dog that is even a little out of control. You owe it to your dog, who by no choice of his own may be judged more harshly than his smaller, less fearsome-looking counterparts, to teach good manners and sociability. Who knows, perhaps your dog can be an ambassador to change perceptions.

To date, five dogs that would be described by many to be "pit bulls" have joined our family: Enzo, Shamsky, Monkey, Lil' Dipper, and Scout. Each has touched our hearts and enriched our lives profoundly. I can't imagine never having had the joy they've all given us.

Misinformation about this breed abounds. Here are some facts you may not know and that you can share with others to spread the good word about these dogs that too often are stigmatized without cause.

FACT: Three breeds of dogs are officially considered pit bull terriers (a mix of bulldog and terrier): Staffordshire bull terriers, American pit bull terriers, and American Staffordshire terriers. Anything else is not a true pit bull, but rather a mix of some other kind.

FACT: Experts have found it impossible to accurately assign a breed label or predict future behaviors based solely on a dog's appearance.

FACT: In the beginning of the twentieth century, pit bull–type dogs were among the most popular family dogs. They were even referred to as "nanny" dogs.

FACT: In the 1980s, pit bulls became the dog of choice of drug dealers, dogfighters, and gangs. These people raise dogs in deplorable conditions and purposely incite fear and aggression in them for their own deviant purposes. The result has been unspeakable injury, suffering, and cruelty for canines and humans.

FACT: The disproportionate number of pit bull types in shelters is more due to irresponsible and prolific breeding than the misconduct of individual dogs. The large number of these dogs in shelters should not be interpreted as a negative reflection on their adoptability.

FACT: Media outlets have always salivated over a good fear-instilling animal story. Some believe that pit bulls became the victim in 1987 when *Sports Illustrated* published a cover with a snarling American pit bull terrier and the headline "Beware of This Dog." *Rolling Stone* also published a graphic article about "teenagers, inner-city gangs, violence, and the horrific abuse of pit bulls." Pit bulls became the new easy breed to plug into the vicious dog story.

FACT: Many shelters euthanize a dog just for looking like a pit bull either due to their own ignorance or because they think that prejudice will limit the dog's chances for adoption regardless of the temperament or soundness of the individual dog.

BLACK CATS:

Shelter Animals with a Bad Rap

SUPERSTITION, BE GONE! Black cats are *not* bad luck! How is it that a superstition that began to form in the Middle Ages in Europe can still have such a strong effect on the opinions we have today in the United States? The majestically beautiful black cat is plagued with a bad rap that continues to haunt it and its chance at finding a good, loving home. It's important for all of us to help shelters by dispelling the myths surrounding these animals and get these deserving sweethearts out of the shelters and into forever homes.

FACT: Black cats are often the last animal to be adopted in a shelter and the first to be euthanized. The same is true for black dogs in a phenomenon known as Black Dog Syndrome. Black and other dark-furred cats and dogs are the proverbial black sheep when it comes to animal adoption.

FACT: The hysteria of witches practicing black magic in the Middle Ages began the negative image of the black cat. It took hundreds of years for this superstition to be put to rest, but the black cat is still looked at negatively by many.

FACT: The genes that cause dark pelts (melanism) also provide a natural sunscreen to the cats. And their black fur helps them to camouflage and hide from predators when needed.

FACT: In many parts of the world, black cats have a stellar reputation. In Italy, when a black cat sneezes, good luck is granted to all who hear it. In England, it is a sign of good fortune if a bride on her way to the altar sees a black cat. In Scotland, if a black cat is found on your porch, it's a sign that money is coming your way. Yes, these, too, are silly superstitions, but they're harmless, positive, and fun.

TEN REASONS TO ADOPT AN OLDIE
(but a Goodie)

SENIORS, LIKE SUSIE of Susie's Senior Dogs, have so much love to give. A lot of seniors were raised in loving homes and have since lost their guardians due to divorce, illness, or death—none of which changed the love between them and their guardians. Seniors are simply looking to share their loving nature with a new family.

1. **They come with few surprises.** There's no need to wonder how big they will grow, how often they will need to be groomed, or what their personality will be like. What you see is what you get!

2. **Bye-bye, potty-training manuals!** Seniors are likely to have already been house-trained—or if they haven't been, they are physically and mentally ready to pick it up in no time.

3. **It's nice to say things just once.** Seniors have been around humans long enough to understand our language. They often know what we are asking or can quickly learn to do as we ask. You can teach an old dog new tricks, and fast!

4. **They fit right in.** A senior dog or cat has been around the block a few times and has come into contact with many other dogs, cats, and people. Seniors usually know what it takes to effortlessly fit in with a family and can do it with ease.

5. **You can relax!** Unlike a puppy or kitten being introduced to a home, a senior animal usually isn't constantly getting into trouble. You don't have to puppy-proof or kitten-proof your house for months on end.

6 **They enjoy brisk walks and don't ask for much.** Older dogs do not require being taken on three runs daily, and they will tire of playing fetch after a short while! Although they do need exercise, seniors are often fine with a nice walk in the morning, aside from potty breaks.

7 **Your favorite new shoes will be safe from doggy damage.** With their teething years behind them, destructive chewing is usually a thing of the past.

8 **Age is just a number.** Age doesn't always mean health problems and expensive medical bills. Young animals can develop health issues as well, and medical bills are usually par for the course throughout an animal's life. Each animal is an individual and deserves to be viewed without judgment.

9 **They give your heartstrings an extra tug.** There is something incredibly powerful about providing sanctuary, love, care, snuggles, and ultimately peace to a senior pet in his or her final years.

10 **Short but sweet time spent together.** Kids go off to college, people retire, and situations change. Sometimes we might have a more limited period of time to devote to the care of a special animal. You can still benefit from the companionship of a super senior.

Susie of the nonprofit organization Susie's Senior Dogs

VIRTUAL ADOPTION:
How It Works

VIRTUAL ADOPTION IS a process in which we advocate for particular shelter animals and give them a voice. In doing so, we hope to find them a forever home outside of the shelter. Through virtual adoption, we get the chance to meet the animals in our local shelter, engage with them, and care for them. If we've got all the furry friends at home that we can handle, or if we are not in a position to adopt, virtual adoption still allows us to take a needy shelter animal under our wing. It gives us the chance to make a difference in the lives of dogs and cats in need, help the busy shelters to care for their animals, and all the while encourage those around us to adopt and not shop. Here's how to do it:

① SELECT AN ANIMAL FOR VIRTUAL ADOPTION

Volunteer your time at your local animal shelter and get to know the dogs and cats there. Choose an animal for virtual adoption, one that you would like to find a good home for. Mutts, pit bull types, black cats, and seniors are all examples of animals that have a hard time finding forever homes. Ask the shelter volunteers about the animal's background and why she has ended up at the shelter. Be inspired to help your chosen friend find her forever home.

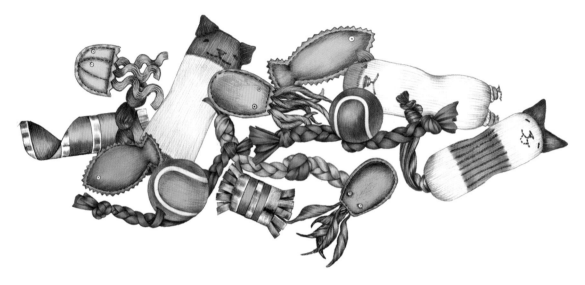

② MAKE AND DONATE ENRICHMENT TOYS

It's important to keep shelter animals mentally and physically active to ensure that no behavioral problems develop while they're in the shelter and to allow them entertainment while they wait to be adopted. Cats can use toys to keep occupied in their cages, and dogs love to be kept busy while they wait to be taken home. Shelter enrichment toys do not need to be fancy or complex. Simply crafting clever items out of recycled materials will suffice and make a world of a difference for the shelter animals.

Host a shelter-crafting day and create a bunch of easy-to-make enrichment toys to donate to your local shelter. (Some of the projects in the preceding chapter might be fun to try.) Don't forget to give some toys to your chosen shelter animal!

3 SPREAD THE WORD ABOUT YOUR SHELTER FRIEND

Posters featuring your virtual adoptee help to spread the word about their need for a good home. Make posters that are colorful and interesting and feature personal details that help potential adopters feel connected to the animal. Add photos and supply all necessary information for the shelter, including directions, a contact number, and the hours of operation.

Cage cards are another clever way to help your animal get adopted. These are the little display notes that are posted on the outside of shelter cages. They provide potential adopters and shelter volunteers with important information about the animal and can include their favorite toys and food, stories about the fun you've had together, and what you believe your animal is looking for in a forever home. They paint a clear picture of who the animal is and the kind of home that would be right.

Simply telling people that you're volunteering at your local shelter and participating in virtual adoption helps spread the word, too. Just by sharing what you're up to, you'll be informing others about the cats and dogs in shelters everywhere who are waiting to be taken to a forever home. Social media has been a godsend to animals in shelters. Instead of posting what you had for dinner, post about an animal in need of adoption!

WANTED

a good home for this pretty gal!

HELLO. MY NAME IS

4 CELEBRATE YOUR GOOD WORK

It's a great day and a great feeling when your shelter animal gets adopted! Before choosing a new animal to advocate for, celebrate your success by finding a way to document the work you've done. Keep a journal or post photos on a bulletin board to commemorate the moment. Make a blog post, update your social media newsfeed, or find another way to spread the word and thank others for their assistance in the project. Talk about it when people ask what you've been up to lately. Who knows, you might inspire your friends to try virtual adoption themselves!

STEWART FAMILY TRADITIONS:
The Foster Charm Bracelet

MY FAMILY HAS fostered many dogs, and we've become much better at it as time has passed. At first, fostering was just a gateway drug to unintended adoption, as those "foster dogs" happily became accidental additions to our lives. Once we'd given an animal a glimpse of what home truly was (a soft bed, a warm body to lie next to, the sweet smell of onions sautéing at dinnertime) and had the joy of watching the animals respond in kind to our love, we just couldn't bring ourselves to return them to the shelter.

Eventually, though, we became more sophisticated in our fostering endeavors and discovered that the real joy of fostering is the challenge of finding other good homes for great pets. We analyzed where true animal lovers hung out—yes, we were on a mission to seek out the most susceptible of the suckers! We'd put small dogs in baby carriers or larger dogs in irresistible T-shirts and off we'd go to horse stables, dog runs, and playgrounds. Within hours, I would be placing a call to my husband, who at first was always expecting to hear that we were once again going to add another member to our family. Instead, I would simply say, "It is done."

Oh, how I love those wonderful, kindhearted suckers!

There are few things in life more satisfying and rewarding than providing a loving home for an incredibly appreciative animal. To all those I've tricked into doing just that . . . you're welcome!

Finding new homes for shelter dogs has become such a part of our family's life that we have developed a nice tradition for documenting our success. We craft small but sweet charm bracelets using Shrinky Dinks that feature all of the animals we have helped. It also works as a conversation starter when people ask about the cute accessory we are wearing. We use it to encourage them to try virtual adoption or, better yet, adopt a new pet!

WHERE ARE THE TROPICAL BIRDS, RODENTS, AND FISH?

IF THEY'RE LUCKY, they are in their natural habitats. (You'll read about some of them in our next section about animals in your backyard.)

I had a reason for including only dogs and cats in this chapter about animals *at home*. Most people are not equipped to provide at home the very specialized and nuanced care that more exotic creatures require. If you are one of those dedicated people who can provide that kind of care and environment to "exotics," as they're called, I hope that you adopt from a shelter and never buy from a store. I also hope that you do lots of research on proper care for your animal.

While we are making progress in protecting the well-being of cats and dogs, we are still painfully behind when it comes to ensuring the welfare of our smaller furry, feathery, and scaly friends. Sold in pet stores, many of these animals are obtained from mill-like facilities that aren't regulated, so there is no incentive for these businesses to spend time or money on humane treatment of their living "stock." If not originating from a mill, a more exotic animal will most likely have been transported from a faraway destination and separated from their animal family. The current rules and regulations for the transport of such animals are designed to protect humans from disease, not to ensure that animals are handled humanely. Regardless of their origin, many don't even make it to the point of sale, as they have died in transit.

Animals raised in squalor or taken from the wild and away from their family are usually distressed and understandably may be inclined to engage in problematic behaviors or may succumb to illness.

Unfortunately, birds, rabbits, mice, and rodents are often considered "starter pets," so many pet stores keep large numbers in stock, making it impossible for even the most well-meaning staff to do an adequate job of providing individualized care and medical attention. These smaller animals are sometimes impulse buys. People bring them home without any prior research into the animals' exact needs. Exotics' sometimes-complex requirements can be foreign to buyers, who are unprepared to meet them. This can lead to the animals' being neglected, returned to the store, or abandoned in the wild; where they can't take care of themselves and starve or fall victim to predators. Make sure you are willing to truly adapt to the special needs of these animals before you make the decision to bring one into your family.

Psst . . . follow me into the backyard.

BACKYARD

Wildlife

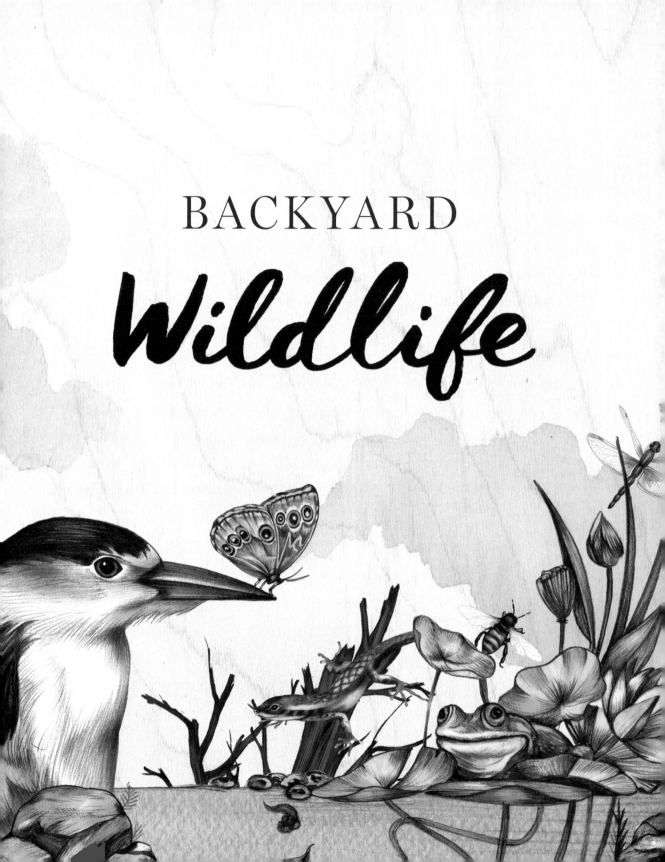

I WAS BORN IN NORTHEAST PHILADELPHIA, which means that my early exposure to animals consisted of my poodle-Pekingese mix, Muffin, and episodes of *Mutual of Omaha's Wild Kingdom*.

When I turned seven, my parents decided it was time to move to the suburbs, and we needed to buy carpets for our new home. We headed to rural Bucks County to shop. I quickly grew tired of the process. Back then we didn't have twenty-four-hour news coverage to terrify my parents into not letting me wander alone behind the store to a creek; instead they hoped I would find some entertainment.

What I did find was my first dead animal: a beautiful starling whose distinctive patterns and colors changed with the light as I circled him. I named him Sparkles because I was seven and the name Muffin was already taken. It was the closest I had ever gotten to anything wild. I was both grief-stricken and exhilarated. I sat beside him and patted his soft head while I prayed that he had found peace.

The funeral was still in progress when my father found me. At first he cautioned me not to touch the bird. Then, finding it was too late for that, he told me we would have to get on the road promptly after I had disinfected my hands. I began to sob. I couldn't leave this poor creature unprotected and alone without a proper good-bye.

My father is a good, good man. He quickly ran back to the store to fetch a shoebox in which Sparkles could lie in state. I cradled the box on my lap the whole way home, trying to cushion him from any bumps. When we got home, we gave our friend a proper good-bye before laying him to rest in our rock garden. In the middle of the night I would sometimes look out to the garden and ask Sparkles to watch over me, as I believed he had gone to a better place where he may have gained more influence over the goings-on of the world.

To practiced suburbanites, the things that are happening in their backyards may seem ordinary. But to someone who has lived life amid asphalt and concrete, a walk in the backyard can be a journey into the exotic.

There are so many valuable lessons we learn from nature—including lessons about patience, compassion, discipline, and fortitude. My family loves to go on nature walks, and I find that some of our most philosophical discussions arise during or after these treks.

Recently on one of our walks, my kids and I found a dead bird. I decided that this was going to be a great learning experience for my eight-year-old daughter and ten-year-old son, so I suggested holding a funeral.

Dead animals can carry many diseases, and the carcasses should not come into contact with your bare skin—but fortunately, I'm one of those weirdos that always has a plastic bag with me. I told my kids that many creatures depend on dead carcasses as their food for the day, so it's important to be sparing and use good judgment when deciding just how many animals we want to hold funerals for. We decided that a funeral was the right thing to do for this little birdie.

We chose a spot that would be easy to visit often and painted a rock to serve as a gravestone. The kids and I all donned rubber gloves before gently removing the body of the bird from the plastic bag and settling it in a small, moss-lined wooden coffin. (Having once run a kids' crafting nature center, I have a supply of wooden craft coffins from past Halloween projects. My ability to instantly produce a ready-made coffin anytime I find a dead animal in the yard can be disconcerting to houseguests.) We talked about what kind of music the bird inspired within us and put together a playlist of songs. We each took turns digging the hole at the burial site while we discussed what we thought the bird's life had been like: What kinds of things did we think this bird had seen? What did we think was this bird's favorite thing to do? Did we think this bird had had lots of friends?

My hope was that this funeral would be part of my larger goal to teach my children respect for the cycle of life and for all living things, no matter how small.

We gathered around the coffin to pay our respects to the bird. I tape-recorded the ceremony because I sensed there would be some gems. My kids didn't disappoint.

Me: You know, guys, this is making me think a lot about my grand-
father who passed away a couple of weeks ago.

My daughter: I was thinking that, too.

Me: Where do you think this bird is now?

My daughter: I think his spirit is all around us now. It's in the flowers, it's
in the trees, and it's in us, in everything, just like Great-grandfather.
The mystery and the gift of life are always worthy of honor.

My Backyard Workforce

WHEN I LIVED in the city, I longed for a backyard. I imagined it would be an oasis for reflection and relaxation. Hey, I might even start meditating. In the summer of 2011, I finally got one—and I quickly discovered that with all the mowing, weeding, pruning, fertilizing, and raking, there wasn't much Zen in my garden. But while I was out there toiling to maintain my private little paradise, I began to realize that I wasn't alone. In fact, close observation revealed a furry, feathery, scaly workforce of garden helpers happily on the job.

The Landscaping Team, Pest Control Team, and Cleanup Crew were out there day in and day out, rain or shine, helping flowers grow, producing fertilizer, eating unwanted bugs, and cleaning up the messes that I preferred to pretend didn't exist. I found I enjoyed taking care of them, and in return, they did my yard right!

The next time you sit back and enjoy the beauty of your healthy, well-tended yard, don't forget to raise your lemonade glass and toast your backyard workforce for their dedication and hard work.

MEET THE LANDSCAPING TEAM

THE LANDSCAPING TEAM visits my garden daily. Bees and butterflies flit about, busily transporting pollen from flower to flower and creating more opportunity to grow. It's all quite sexy. Their whirring wings and colorful uniforms brighten the day and make everything seem right with the world. Worms, squirrels, rabbits, and moles are also hard at work spreading seeds along the grass, eating pesky grubs aboveground and below, and digging tunnels that provide oxygen to deep layers of soil, making the soil healthier—less sexy, but still necessary.

The Landscaping Team plays a crucial role in our garden, and it's important that we not unintentionally harm team members while they work. Often people are unaware of the effects of particular garden and lawn products on these animals. The numbers of bees and butterflies are decreasing due to the destruction of their habitats by the pesticides we spray, some fertilizers containing harmful chemicals inflict damage on earthworms, and certain methods of pest control lead to unnecessary disease, injury, or death for rabbits, squirrels, and moles.

We have a responsibility to keep these hardworking and innocent creatures safe and healthy. Once we are aware of what they do and what they need us to do in order to help them perform their jobs, we can learn to live in harmony and implement natural and humane methods of both encouraging and, when necessary, deterring these animals from entering our yard. In my backyard, we think of ourselves as a team with a shared habitat.

THE LANDSCAPERS:
What Is Everyone Working On?

THE BEE

There are a number of different bee types that pollinate flowers by collecting nectar for honey-making or pollen for feeding their young. As they fly and land on different flowers, pollen spreads and allows plants to reproduce. Bee pollination is responsible for almost half of the food we eat, so maybe next time you find yourself fleeing a bee, you should at least yell back, "Thanks!"

THE BUTTERFLY

Beautiful, beautiful butterflies collect and spread pollen when visiting flowers on their hunt for food. Butterflies land feetfirst on the flower, using their feet as taste receptors. Pollen attached to their feet falls onto other flowers and pollinates them. Butterflies' color vision gives them an advantage over bees, as it allows them to spot some flowers that bees may have missed.

THE EARTHWORM

Poor little earthworm! Worms' sliminess can be so off-putting, yet they work so hard underground to keep our gardens healthy. The earthworm wriggles through the dirt and soil, allowing air to reach the roots of plants. They excrete healthy soil fertilizer after spending all day eating decaying leaves, fallen fruits, and kitchen scraps they're lucky enough to find.

THE GRAY SQUIRREL

The fluffy-tailed gray squirrel collects nuts and seeds for food and buries them underground, storing them for the colder months. Fortunately for us, squirrels often forget how many seeds they've buried and exactly where they've hidden them. The abandoned seeds eventually sprout into trees. The squirrel is unwittingly responsible for replenishing much of our ecosystem and helping hundreds of new trees to grow each year. If only our own forgetfulness could be so fruitful!

THE RABBIT

It's hard to get too mad at a bunny, but they can be thieves in our gardens. They do, however, leave behind some useful gifts once they've visited. Rabbit droppings contain large amounts of nutrients that help fruits and flowers grow, and rabbits' underground burrows, often considered nuisances by gardeners, are beneficial, as the holes allow for oxygen, sunlight, and water to reach the roots of certain plants.

THE MOLE

Unfortunately, in real life, moles don't wear adorable spectacles as they do in cartoons. If they did, gardeners might not get so upset about their underground tunnels, which produce large mounds that disrupt the look and layout of lawns, flower beds, and vegetable gardens. However, as underground travelers, moles do find harmful insects and grubs that eat the roots of our plants and cause damage to our garden's growth.

POLLINATOR GARDEN:
What I Plant to Create Buzz

MY POLLINATING TEAM includes bees, butterflies, humming-birds, bats, moths, and—all right—even flies. Pollinators are some of the most important creatures in my yard. Although they are small, without them I would not have such a healthy, fruitful garden. Different pollinators are attracted to and able to see different colors. Adult butterflies are attracted to white, pink, purple, red, yellow, and orange flowers. Blue and green flowers are butterflies' least favorite. Bees can see yellow, blue-green, blue, and ultraviolet. They cannot see red or green and are warned off by black. Planting native flowers is the best option for keeping any garden healthy, and growing an array of species will make sure to keep all of the pollinators busy, happy, and frequent visitors. When you are planning your garden, take some time to find out what grows natively in your region, and serve up a menu of colors and shapes to delight local pollinators' visual palates! Below are some flowers that are native to my garden and help to attract all kinds of pollinators.

black-eyed Susan

wild columbine

goldenrods

salvia

MY WILD, BEAUTIFUL YARD:

Sometimes the Grass Is Greener When It's Not So Green

SOME VIEW WILDLIFE as the nemesis of their manicured lawn, but I see it as the best part. Before calling in the exterminator, think about the repercussions, not only for the animals but, in the case of poisons, for ourselves. Not all pesticides are bad, but many can affect our nervous system, and others may irritate the skin or eyes. Some pesticides may be carcinogens, and others may affect our hormonal or endocrine system. It's important to ask lots of questions before letting loose with pesticides. When I looked out my window one day and saw a man wearing a mask spraying the trees near my home, I didn't sit idly by and assume it must be okay. This was where my kids would roll around in the grass, where my dogs would sunbathe, and where my bunnies would dine.

Personally, I'd always rather explore options and try a more natural, less toxic, nonharmful product or technique first. They do exist. I'm also woman enough not to care if my neighbors judge me for not having a "perfect" lawn. Our natural world can do some pretty miraculous things when it's supported and appreciated.

fairy candles

milkweed

woodland pinkroot

turtle heads

BE NICE
TO THE BUGS!

WHERE HAVE ALL THE BUTTERFLIES GONE?

The most recognizable butterfly in North America is the monarch: the orange-and-black-winged creature that disappears in the late summer months. But where does the monarch butterfly go?

Monarch butterflies migrate from the United States to Mexico once a year to escape the cold of winter. As creatures that prefer humid temperatures, they must migrate to the warm southern climate for survival. In order to encourage them to return in the spring, we can plant milkweed in our pollinator gardens. Milkweed is the host plant that monarch butterflies use to lay their eggs. Unfortunately, milkweed is becoming harder to come by due to the large use of pesticides and deforestation in the United States and Mexico. A decrease in milkweed means a decrease in monarch butterflies. It's up to us to help the monarch butterfly escape near endangerment by encouraging the growth of milkweed in our yards so these useful creatures have a place to grow, live, pollinate, and survive.

WHY ARE THE BEES IN TROUBLE?

All species of bees—honeybees, mason bees, carpenter bees, and more—rely on flowers and flower-rich grasslands to survive. With traditional agricultural practices being replaced by industrial agricultural practices, wildflowers and grasslands are being wiped out, sprayed with chemicals, and overtaken by factories.

MY BACKYARD WORKFORCE

All of these changes to the environment make it close to impossible for bees to gain access to flowers, and therefore they struggle to survive. This is a problem for humans, other animal species, and nature itself, as the bees are responsible for much of the food we eat and for the foliage that surrounds us.

We can help the bees by planting more flowers to attract them to our gardens and supporting small apiaries that are working to keep the bee populations alive.

MY PESTICIDE-FREE GARDEN SPRAY RECIPE

I have a lot of fun planting native flowers in my garden to attract local bees, butterflies, and birds. I use natural soil and fertilizer and keep my garden pesticide-free by making a homemade garlic spray. Here's the recipe.

- Peel and crush two whole bulbs of garlic into small pieces.
- Place the garlic pieces in a bowl and cover with enough boiling water to fill your spray bottle.
- Cover the bowl and let the mixture steep overnight.
- Strain the mixture to remove garlic pieces, and pour the liquid into a spray bottle for use on plants.

I usually spray my garden once a week if it hasn't been raining, concentrating on the base of the leaves, where I know the critters love to munch.

The little insects and bugs flying around are all part of nature, and though they might eat away at some leaves, I leave them alone, because I know there is another animal out there looking for *them* as a meal. Everybody's gotta eat!

A HOME FOR THE MASON BEES

MANY BEES BUZZ around the garden, but the solitary bee is the most effective bee to encourage in our backyard. Solitary bees, like the mason bee, live on their own and not in colonies with a queen and workers (like the honeybees and bumblebees do). The solitary bee is her own queen. She builds her own nest, collects her own pollen and nectar, and lays her own eggs without any help from other bees. She is also very unlikely to sting you. As the bee population continues to decline, encouraging all bee types is crucial.

Make

BEE HOUSE

We can provide a place for the harmless mason bee to lay individual nests for their larvae. They are attracted to all kinds of materials for nest-making, from bamboo stems to blocks of wood, hollow logs, and hollow sticks, as well as store-bought nesting tubes made of paper or of natural reeds.

SUPPLIES

- Carpenter's glue or hammer and nails
- Wooden or timber box, 8 inches deep and open-faced
- Flat piece of wood or timber 2 inches bigger than the size of your box, for making a slightly overhanging roof
- Saw or other cutting instrument
- Bamboo stems or store-bought mason bee nesting tubes (enough to fill the box)
- Plumber's strap (available at any hardware store) or wire for hanging the house

DIRECTIONS

1 Glue or nail the flat piece of wood to one short side of the wooden box to create a roof. Be sure that it overhangs the edge of the box slightly to protect the house from rain.

2 Cut the bamboo stems 8 inches long, to match the depth of the wooden box.

3 Place the bamboo stems or nesting tubes in the box with the open hole on one end facing outward. Be sure to fill the entire space.

4 Choose a spot to hang the bee house. You should choose an area that will receive direct sunlight for part of the day but not get too wet on rainy days. If the bamboo stems or wooden box become too wet, they will rot and become useless. The house should be at least 3 feet off the ground and free of vegetation that may block entrance to the bamboo stems or tubes, where the bees will lay their eggs.

FEED THE WORKERS BELOWGROUND:

The Earthworm

EARTHWORMS DIG TUNNELS in the ground, which aerates the soil. They also produce fertilizer that keeps our land healthy and alive. As hard workers in the garden, they deserve a treat. We can give back to the earthworm by offering some of their favorite snacks to make it easier for them to do their job.

I do my fair share of filling landfills with things that will never properly decompose. In an attempt to make up for that, I deposit our kitchen waste in a worm bin. The worms happily eat the scraps and turn them into fantastic compost, which I then use on my plants. It's a win-win!

The kids love lifting up the lid of our bin, spying on the worms, and seeing how much they have eaten each day, and I like knowing that these little creatures are being fed. It's the mom in me.

Make

WORM BIN

SUPPLIES

- Kitchen scraps and other items worms like to eat (see list)
- Container/bucket around 16 by 14 inches and 8 inches deep (recycle an old drawer, wooden box, or plastic bin)
- Dirt (potting soil or soil from your garden)
- 1-inch-wide damp newspaper strips and leaf litter
- Worms (I like to send the kids out to collect them from the yard)

DIRECTIONS

1   Cut food scraps down to a small size to make them easier for worms to eat.

2   Add scraps to the container, layering the scraps with moist dirt, newspaper strips, and leaf litter.

3   Place the worms gently into the container and allow the feeding to begin.

4   Stick to a 2:1 ration—for every 2 pounds of worms, give 1 pound of food. Only add new and fresh scraps when the supply looks like it needs to be topped up.

5   Keep your worm bin in a cool, dark area.

GOOD FOOD FOR EARTHWORMS

- Potato peels
- Corncobs
- Lettuce
- Melon rinds
- Apple cores
- Banana peels
- Eggshells
- Coffee grounds
- Tea bags
- Wet cardboard (avoid cardboard that has been marked with ink)

Note: Do not give earthworms meat products, bones, dairy products, or anything too acidic, such as pineapple, onion, or citrus fruits. These items take much longer for the worms to digest and break down and will cause your bin to smell and attract pests. Salty foods should also be avoided, as they can dehydrate worms and cause them to die. (I also try to keep these guidelines in mind when it comes to what I put in that other waste bin known as my belly.)

FEED THE WORKERS ABOVEGROUND:
The Squirrel

SQUIRRELS DIG HOLES and bury their stash of nuts and seeds, which eventually encourages trees to replenish and grow. All of the work these critters do, and the impact it has on our world, comes from their desire to eat, and eat a lot!

Squirrels used to attack my outdoor bird feeders before I gave them a feeding place of their own with this easy-to-make feeder. My kids and I love to spot a squirrel perching on the fence where we've secured our little jar of treasures. We watch them eat, and at the end of every day, we make sure that the jar is empty of food so that other nocturnal creatures, like raccoons and opossums, don't ransack the feeder.

Make

SQUIRREL FEEDER

SUPPLIES

- 2-by-12-inch flat wooden stick
- Widemouthed glass jar
- Twine or wire
- Squirrel food (see list)

DIRECTIONS

1 Attach the stick to the jar, using twine or wire to secure it tightly at the top and bottom. Make sure that the stick extends 5 inches or more beyond the jar opening to act as a plank for the squirrel to perch on while eating.

2 Fill the jar with treats that squirrels love best.

3 Attach the feeder with wire to a stable object, such as a fence railing, in a spot where the squirrels are likely to find it.

GOOD FOOD FOR SQUIRRELS

- Sunflower seeds*
- Walnuts*
- Hazelnuts*
- Shelled peanuts*

* All nuts and seeds should be raw and unsalted

LIVING IN HARMONY WITH OUR BURROWING BUDDIES

MY DAUGHTER AND I are a rabbit's idea of perfect gardeners. My love for the book *The Secret Garden* led me to build a garden for my daughter in our yard. I imagined picking all sorts of vegetables and returning to the kitchen to create a feast from our bounty. But our harvest turned out to be greater than my desire or ability to cook! Fortunately, three rescue bunnies had just come into our lives. My family certainly enjoys our homegrown healthy vegetables, but we're also more than happy to share them with the rabbits. Of course, if you'd prefer to keep your garden to yourself, there are some humane ways to live in harmony with burrowing landscapers and deter them from munching on your plants without the use of poisonous chemicals.

1. Rabbit-resistant trees, shrubs, and bulbs can be grown in the garden. Examples include oak trees, pine trees, cedars, Oriental poppies, butterfly bush, boxwood, and many more.

2. Lavender plants help to deter rabbits, as do sage and rosemary.

3. Twigs and leaves scattered on open grass can help distract burrowers from plants and trees.

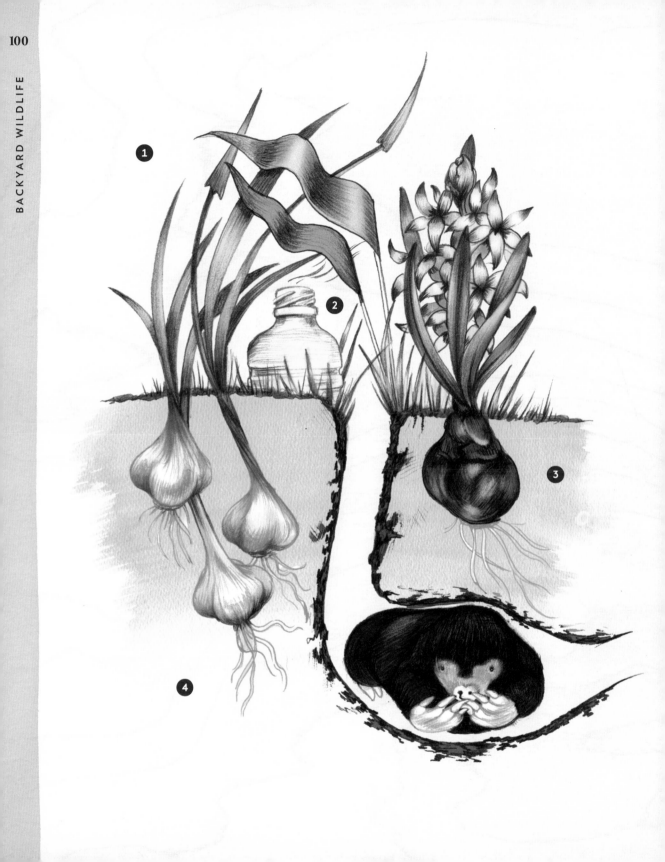

THE COST OF OTHER METHODS

Flooding and trapping moles underground has proven to be an ineffective solution for keeping them out of the garden and can cause the animals to contract diseases, injure themselves, or suffocate and die. A mole that has been trapped underground can be injured while attempting to escape or simply as a result of the trapping itself. An injured (and diseased) animal can cause disease to spread throughout the yard, affecting other animals, household pets, and gardeners who attempt to remove the carcass. Trapping rabbits can have a similar result. Other methods of deterring rabbits, including the use of ammonia-soaked rags left along a rabbit-frequented areas, can be just as harmful. Often recommended as a "convenient household product" with an off-putting scent, ammonia can damage a rabbit's lungs and result in pneumonia and other infections. These methods are cruel and do little other than cause innocent animals to suffer and eventually fall victim to their symptoms. Most of these control attempts simply result in a temporary vacancy that will be filled by other creatures.

1 Humane scare tactics can frighten animals away. Strips of reflective tape hanging from trees or flags placed around the yard can scare off rabbits and moles with their mirrored reflections and fluttering sound.

2 An empty bottle left at the entrance of a mole's nest, right side up, can set off a wind-whistling noise that will prompt moles to move elsewhere.

3 Moles are repelled by the sweet, lingering fragrance of hyacinth bulbs.

4 Watering the garden helps to keep moles away, as moles dislike wet soil.

MEET THE PEST CONTROL TEAM

MEMBERS OF YOUR Pest Control Team help ensure the balance of nature in your yard. They keep your garden from being overrun by plant-destroying critters by roaming the yard in search of their daily meal. Some work the night shift and even double as landscapers: bats and birds of various species visit the yard when the sun goes down and fly among the fruit trees and flowers, spreading seeds and helping them to flourish. I am particularly fond of the pest controllers that keep mosquitoes out of my yard. Frogs, spiders, reptiles, and insects are doing valuable work out there.

There are simple techniques to welcome these pest controllers into your yard and promote their good work, just as there are easy and gentle methods to deter them if their presence becomes unwelcome. It is important to remember that once you begin to welcome animals into your yard and encourage them to visit a food source or shelter option, you must be consistent with your offerings. Once they have become aware that food and shelter are available in a particular place, they will begin to rely on it.

THE PEST CONTROL TEAM

What Is Everyone Working On?

THE FROG

Frogs live in ponds and wet areas around our gardens. They eat decaying leaves and keep the surrounding areas free of brown foliage. In these moist environments, they are surrounded by mosquitoes and sow bugs, which they capture using their long, sticky tongues. Spotting frogs in your garden is an indicator of a healthy environment.

THE SPIDER

Don't squish that spider! The spider catches and eats insects that would otherwise hang around and destroy our flowers and leaves. Some spider species prey directly on insects, while others use the more passive approach of spinning sticky webs to ensnare the unwary.

THE REPTILES

Snakes, lizards, and turtles feed on many harmful bugs and insects, as well as on mice and rats. Ninety-nine percent of the snakes we find in our gardens are harmless helpers that are beneficial to the way our healthy garden works.

THE BIRD

Many birds visit our gardens when birdseed and attractive plants are available in the yard. Their visits help keep our garden healthy and clean. Swallows and swifts catch insects as food for themselves and for their young. Finches and sparrows eat large quantities of weed seeds that help limit the growth of unwanted plants on our lawn.

THE BUGS AND BEETLES

While some bugs and beetles do damage to our garden, others are actually beneficial. Ladybugs feed on insects from the minute they are born. A single ladybug can eat up to 5,000 aphids in its lifetime, helping us to protect our crops and plants. Gardeners often release a group of ladybugs into the yard for a coordinated attack by this elite bug brigade. Ground beetles are nocturnal predators of slugs, snails, and other creatures that live in our soil. And roly-poly bugs (more formally known as pill bugs) survive on water-soaked materials such as carrion and rotting vegetation, making them useful cleaners in our yard.

THE BAT

Mosquitoes love me, so I love bats! The bat is a nocturnal creature that works the night shift, eating garden pests and pollinating our plants. As bats visit fruit trees for midnight feeding, they pick up and drop off seeds and fruit remnants, allowing these bits and pieces to take root and eventually flourish into new trees. One bat can eat 2,000 to 6,000 insects per night, including 600 mosquitoes in just one hour. Bat droppings also work as fertilizer for our soil, keeping plants flourishing.

BACKYARD BIRD B & B

BIRDS HAVE MIGRATION routes that they follow each year, but due to overdevelopment and loss of habitats, many local feeding spots along their routes are no longer in existence. As someone who appreciates well-appointed accommodations on my travels, I take great pride in creating my backyard bird B & Bs so the birds who pass through our yard always get a delicious meal and comfortable quarters. They return my kindness by visiting my backyard from year to year.

A backyard bird B & B should be a sanctuary for birds. The best B & Bs are those with a reliably delicious menu, a comfortable rest area, and a feeling of safety in a home away from home.

I fill feeders with a variety of seeds to meet the different preferences of each of my bird guests. As I've mentioned, just like our own pets, animals in the wild become accustomed to the food we provide for them. I attempt to be consistent in what I offer. I also make sure to put window stickers on any glass doors, helping birds identify a clear flight path and keeping them out of danger.

Make

SIMPLE BIRD FEEDER

SUPPLIES

- Scissors or box cutter
- Empty plastic bottle
- Wooden spoon (make sure the handles are longer than the diameter of the bottle)
- Sisal twine
- Store-bought or homemade birdseed (recipes follow)

DIRECTIONS

1   Cut a ½-inch circular hole 4 inches from the bottom of the bottle and a 1-inch circular hole on the opposite side.

2   Insert the spoon handle first through the 1-inch hole so that the handle protrudes from the ½-inch hole and the bowl of the spoon protrudes from the 1-inch hole.

3   Remove the bottle cap and fill the feeder with birdseed.

4   Use scissors or a box cutter to carefully pierce a hole in the middle of the bottle cap.

5   Thread a piece of twine through the hole and secure it by tying a knot at the end that's inside the cap.

6   Screw the bottle cap back onto the bottle and use the twine to hang the bird feeder in a tree, by the porch, or anywhere in the yard where it will be accessible to the birds but out of reach of squirrels and other backyard animals.

KEEP YOUR BIRDS SAFE

THERE ARE THREE simple methods you can implement to keep your backyard birds safe and comfortable.

1. **Use window stickers.** Buy simple window decals or make your own stickers to apply to your clear windows and doors. This will stop birds (and other animals) from flying into the glass and injuring themselves.

2. **Keep cats indoors.** Domesticated cats should very rarely, if ever, be allowed outside in the yard, as their natural instincts will kick in and they can and probably will attack birds and other wild animals. This can cause harm to both your cat and the wildlife.

3. **Hang a squirrel protector.** Squirrels and other small mammals will desperately try to get their hands on your backyard bird feeder. Keep them away by hanging a store-bought squirrel protector or adding an upside-down plastic bowl to the top of your feeder to prevent squirrels from gaining access to the food.

Make

RECIPES FOR YOUR FEEDER

Different bird species prefer a range of ingredients in their feeders. Attract these common birds using the following bird food recipes.

NORTHERN CARDINAL

- Sunflower seeds, crushed peanuts, white bread, raisins, bananas

WOODPECKER

- Sunflower seeds, crushed peanuts, apples, oranges, melon seeds

BLACK-CAPPED CHICKADEE

- Sunflower seeds, crushed peanuts, piecrust

MOURNING DOVE

- Sunflower seeds, bread crumbs, apples, bananas, pears

A MOST SOOTHING HOBBY:

Tips for Bird-Watching

- **Use binoculars.** Binoculars can help you to see more clearly when bird-watching.

- **Bring a camera.** You might get some great shots not only of the birds but also of the watchers. (We always get some nice family shots.)

- **Bring a field guide.** A field guide is packed with information and will help you to identify the various birds you count. We look through the guide for a couple of days prior to our actual outing to familiarize ourselves with what birds we are most likely to encounter. We appoint one member of our family as the field guide looker-upper so the rest of us can keep our eyes peeled for birdies.

- **If you love to sketch, bring a notepad and colored pencils.** Record your findings and make sketches to help keep track of and remember what you've seen. Pay attention to any distinctive stripes or color patches, eye rings, crests on the head, shapes and colors of wings and tails, and markings on the body, including under the belly and along the tail.

- **Keep your eye on the bird for as long as it sits still,** and write down everything you noticed once it has flown away or is out of sight. It's a great exercise to improve your powers of observation!

- **Listen for songs.** The song of a bird is one of the best ways to identify its species. Keep an ear out for the bird's vocalizations and try to tune in to the song it's singing.

- **Flock together.** When we go out as a family, we stay together and do our bird count as if we were one unit, so as not to duplicate any sightings. But if you make a fun winter party out of it, everyone can split up and cover a lot more ground. Invite friends over; fan out in different parts of the backyard or woods; count, draw, and photograph to your hearts' content; then come back to the house to flock together and swap stories over a cozy meal.

HOW TO HUMANELY DEAL WITH UNWANTED BIRDS

If birds are roosting in your garden and you'd prefer they didn't, choose humane methods of keeping them out. Visit your local garden supply or hardware store and ask for the appropriate netting or wiring to use to stop birds from congregating in areas where they are not wanted. If you find birds hovering around your vegetable garden or pecking at your fruit trees, try purchasing a soft, no-tangle, 5-inch mesh to cover the garden or trees and keep the birds out of reach of your produce. Poisoning birds and using Avitrol or other avicides as dispersing agents are cruel and ineffective ways to resolve problems with birds. They can impair a bird's nervous system and cause birds to display erratic behavior and suffer convulsions before eventually succumbing to the effects of the toxins. They will not stop birds from roosting in your yard in the future and are only a short-term and brutal solution. If you see a bird that is showing signs of poisoning, call your local wildlife rehabilitator or veterinarian immediately.

Make

A FROG SANCTUARY

A neighborhood that is native to frogs is a great place to grow a garden, and supplying a sanctuary for them on your property is the best way to keep them around. Often, frogs won't live in your pond, spending most of their time in the surrounding plants, but they'll use it to breed, and their babies, the tadpoles, will live in the pond. No nonnative frogs should be forced to live in an unfamiliar pond, as they will not be able to survive in conditions unnatural to them. Be sure to build your pond at least 30 feet from your house (and your neighbors' houses); otherwise the frogs' croaking might keep you up at night!

SUPPLIES

- Shovel
- Plastic or rubber pond liner (available at stores like Home Depot)
- Sand
- Rocks and logs of various sizes
- Garden pots (sized to fit the emergent and submergent plants)

- Submergent plants, such as milfoil (ask your local garden store for plants native to your area)
- Garden topsoil (available at any garden store)
- Gravel pieces
- Emergent plants, such as bulrushes and water lilies (ask your local garden store for plants native to your area)
- Floating plants, such as duckweed (ask your local garden store for plants native to your area)

DIRECTIONS

1   Dig a hole in the ground that is half the size of the pond liner you have purchased. A good size for a frog pond is 12 feet long by 6 feet wide by 2 to 3 feet deep.

2   Remove any rocks or roots that might pierce a hole in your liner, and use the dirt you've dug up to build sloping sides and banks on the pond for easy access by the frogs for getting into and out of the water.

3   Line the hole with a layer of sand approximately 2 inches deep.

4   Spread the pond liner in the hole, allowing it to extend over the hole by 12 inches all around.

5   Fill the hole with water. Allow the water to sit for a few days before adding plants; this allows any chlorine in the water to evaporate.

6   Pot the submergent plants using a good garden topsoil (not potting mix or peat moss, as they are too light and will become loose in the pond). Cover the surface of the soil in the pots with gravel to weigh the pots down, and add them to your pond.

7   Pot the emergent plants using a good garden topsoil as well. Cover the surface of the soil in the pots with gravel and add them to the pond.

8   Set your floating plants in the water.

9   All plants in and around the pond should be native and should cover 50 to 70 percent of the pond. Long grass around the pond is also effective for attracting insects for frogs to feed on.

DON'T PESTER THE "PESTS" THAT DON'T PESTER!

BATS, SPIDERS, REPTILES, and bugs are plagued with a bad rap and are undeservedly considered "pests" by many people. Rather than be reviled for the misconstrued myths surrounding them, these animals deserve to be praised for all of their great qualities.

1. Bat droppings are called "guano" and are one of the richest fertilizers for soil. Bats pollinate as they fly, making them responsible for the growth of fruit trees in many different regions.

2. Some spiderwebs are strategically spun and positioned by spiders expecting to trap their prey, which includes pesky insects that eat our plants. Sometimes the webs can capture mice or even other spiders.

3. Ground beetles spend the majority of their time searching for prey, which includes caterpillars, slugs, and snails, all of which may chomp holes in our garden leaves.

4. Some snakes and lizards munch on plants, but most survive on insects such as ants, grasshoppers, wasps, and beetles (many of which are nuisances in our garden).

MEET THE CLEANUP CREW

THE CLEANUP CREW scavenges around our backyard and front lawn, optimistically looking for food to eat. These wild animals were technically living on our grounds before we were, so they should not be considered trespassers! It's their home, too. As we humans move farther and farther onto their land and their natural habitats, so, too, do they move into ours. Raccoons, opossums, and skunks are merely taking advantage of the fact that since humans have come to town, there is a greater abundance of food available, and more variety at that!

While scavenging isn't always a highly admired quality in the human world, these animals are simply following their natural instincts—and their scavenging ways offer a special kind of help to us. Animals such as owls eat insects and rodents that would otherwise be feeding on our plants and living as a nuisance in our lives. Crows, foxes, opossums, and skunks all take charge of roadkill, cleaning up these messes that we prefer not to touch. They help our ecosystem by breaking down organic materials and recycling them into nutrients needed for growth in the environment. They venture into our urban environments and feast on garbage scraps that we discard, and since they're not accustomed to the contents of trash cans and the dangers of cities, their scavenging in these places can actually cause more trouble for them than it can for us. We must be aware of the situations that innocent animals can find themselves in and do what we can to keep them from harm.

Sure, we'd prefer that they not take up residence in our chimneys or rooftops, but there are humane ways around those situations and smart solutions to living in harmony. We can keep these animals at a distance and still appreciate them for how they work to keep balance in our world.

THE CLEANUP CREW

What Is Everyone Working On?

THE RACCOON

With their facial markings that make them look as if they're wearing black masks, raccoons are the "bandits" of the animal kingdom. They sneak around at night and can get a reputation for pawing through our trash cans, crashing around, and making noise. Raccoons have nimble fingers and can unlatch cages and unlace strings. They help us by cleaning up roadkill and other carcasses that we do not want to see or touch.

THE OPOSSUM

Another night scavenger, the opossum roams our lawns and invades our garbage cans in search of rotten fruit and other food. Opossums are often seen cleaning up the roadside, eating roadkill and food-scrap litter left by untidy humans.

THE CROW

The crow will eat almost anything, including animal decay and garbage. Crows are one of the smartest animals in our animal kingdom and can memorize the routine of garbage collectors, working out the best time to visit each spot in order to ensure that they can swoop down for any leftover scraps and snacks. Crows clean up roadkill and help control the populations of insects and other small creatures.

THE OWL

The owl is a nocturnal hunter that preys on everything from insects and bugs to small mammals and fish. Owls can fly almost silently, enabling them to surprise their prey in the dead of night. This, coupled with their strong vision and specialized hearing, makes their hunting capabilities stronger than those of most other birds.

THE SKUNK

The skunk eats carcasses and pesky insects at sunrise and sunset. Their predators stay away as they know about skunks' nasty-smelling spray. Skunks will release spray if they feel threatened, stamping their feet, raising their tails, and hunching their backs to warn that a spray is on the way.

THE RED FOX

Foxes are the most opportunistic feeders and eat up to 2 pounds of food every day. The sly red fox has the ability to hunt even during the snowy winter months, helping our environment to stay balanced while the other Cleanup Crew workers are hibernating. They also help to clean our gardens by hunting down rodents, insects, beetles, and unwanted plants.

THE CROW:

Who Are You Calling a Birdbrain?

T HE CROW HAS been recognized as one of the smartest creatures in our animal kingdom. A murder of crows (yes, that's the official term used to describe a group of crows) uses its own dialect to communicate. They are incredibly social animals that live in family groups and remain loyal and protective of any other crow species or family nearby. If one crow lets out a loud distress call, others will rush over to help and defend. And if you've ever pulled a nasty stunt on a crow in the past, don't fool yourself into thinking that the crow has "moved on"! Crows recognize a human face for a lifetime and will remember those who have caused harm or danger to them or their family. It's best to steer clear of them; you don't want to be on a crow's hit list!

One of the crow's most brilliant qualities is his ability to solve intricate puzzles and use everyday objects as tools. Twigs, grass stems, wire, and leaves have all been used by crows to meet their needs and help them access places they are determined to get to. Pieces of wood have been used as probes, while twigs and grass stems have been bent and plucked to reach food stuck in logs or between rocks. Some crows even use sticks as "fishing" tools: they sharpen the edges, poke and prod in the water, and eventually attract a bite from an animal that is then drawn out of the water and eaten for supper.

THE OWLS:
My Childhood Idols

WHEN MY YOUNGER sister and brother were born eleven months apart, I decided it was time for me to spend more time at my grandparents' house. They were devoted nature lovers, and being with them was like visiting a pair of forest elves—even though they lived in Philly! We'd fill bird feeders in the yard, rub peanut butter on large rocks and pinecones for the squirrels, and eagerly wait for all of the hungry critters to show up.

Even inside the house, wildlife was everywhere. Every painting, photograph, and decorative piece was of an animal. My grandmother loved owls, and over the bed that I slept in was a collection of owl photographs, paintings, needlepoints, and drawings. As I readied myself for my nightly prayers, I'd stare up at them. I was in parochial school at the time, so God had mostly been represented to me as a handsome gray-bearded fellow in a flowing white caftan. But as I looked into the wise eyes of the owls, I felt convinced that long ago God must have taken up residence in these beautiful, regal birds. Under their watchful eyes, I felt safe and protected. I'm still not convinced I wasn't onto something.

(It wasn't until I wrote this story about my grandmother's home that I realized my home today could be described in exactly the same way. I even still go to sleep next to a beautiful sculpture of a barn owl that sits watching on my nightstand. Thanks, Grandma!)

THERE'S BEAUTY IN THE NIGHT

We all know owls are wise, don't like to litter, and want to get to the center of a Tootsie Pop, but owls are not all the same.

WESTERN SCREECH OWL

The urban owl. His brilliant branch-stub disguise allows him to camouflage with ease.

BARN OWL

The heart-shaped-face owl. This ghostly pale owl is strictly nocturnal and masters the art of silent flying.

GREAT HORNED OWL

The quintessential storybook owl. His earlike tufts and deep "hoot" intimidate his predators.

SPECTACLED OWL

The spectacle-wearing owl. The white lining surrounding his eyes gives him both his distinguished appearance and his name.

BOREAL OWL

The deep forest owl. With asymmetrical ear openings, this owl has a keen sense of sound and can gauge both height and distance at once.

SNOWY OWL

The winter white owl. With white plumage and small black markings, this owl can keep hidden day or night.

BURROWING OWL

The underground owl. His desire to inhabit burrowed holes and live a diurnal life makes him one of a kind.

ELF OWL

The smallest owl. When scared, this tiny owl plays dead until all danger has passed.

NORTHERN SAW-WHET OWL

The tiny but tough owl. This very small owl has an oversized head and a strong sense of dominance.

THE SCAVENGERS' TRADEMARK SKILLS

EACH MEMBER OF the Cleanup Crew has a sneaky skill that gives it an edge over its predators. Take a look at some of their secret abilities and get to know the tricks they play on the other creatures in the wild.

THE THRIFTY BANDIT

The raccoon's nimble fingers, even without an opposable thumb, are agile enough to twist, turn, unlatch, and pull open objects such as garbage cans, refrigerators, doorknobs, and container lids. Their black-fur eye mask is the perfect costume for these dexterous creatures that sneak in at night and raid for goodies.

THE DRAMA QUEEN

The opossum's ploy of playing dead to escape danger has become so well-known that pretending to be something we are not is often called "playing possum." Opossums "play possum" when caught out in the daylight with little chance of escape. They drop to their sides, curl their bodies, drool out of their open mouths, and excrete droppings to give the appearance of being dead. They can maintain this pose anywhere from minutes to hours, until they decide when to wiggle their ears and listen for sounds to determine whether or not it's safe to get up and move on.

THE CUNNING DETECTIVE

The red fox possesses the ability to hear low-frequency sounds, both aboveground and below. Like detectives, these animals use their heightened sense of hearing to locate prey and then deploy their trademark four-step hunting strategy: a quiet and careful move toward the sound of the prey, a freeze in motion to confuse the prey with unpredictability, followed by a leap and a pin to the ground.

THE STINK BOMBER

The skunk's not-so-secret skill is, of course, emitting its powerful spray, which is used as a predator deterrent. The spray is produced from glands underneath the skunk's tail and can be sprayed as far as 10 feet, which keeps predators at a good distance. While the spray is not poisonous and doesn't cause any real damage, it is extremely potent, lingering, and uncomfortable on the predator's skin, and the odor is almost impossible to get rid of.

A Mindful Nature Walk

JUST BY WALKING outside, we have the opportunity to observe different creatures in their natural habitats and take in all that they do from day to day. Some creatures have adapted to life around humans and are willing to come a little closer, while others prefer to keep their distance. It's a great way to get outside and enjoy the amazing animals that surround our homes without disrupting their environment or well-being.

Don't think that if you're a city dweller, you can't discover wildlife. Look around! Animals abound!

THE HURTLESS HUNT:
How It Works

I T'S NOT HARD to have a Hurtless Hunt—there are just four steps. You can do a solo Hurtless Hunt and connect with nature in a whole new way. Or try it with your mate, your kids (and maybe a few of their friends—but not too many, since noise will scare away the critters), or some friends of yours. You might stay in your own backyard, or you might venture into nearby woods, fields, water areas, or parks. The trick is to get quiet, tune in, and open your eyes and ears. I guarantee that the Hurtless Hunt will open your heart to the many creatures living and working all around you. If you really pay attention, you're sure to come up with a way you can help them.

1 PACK YOUR BAG

I have my kids pack a bag with items they'll need to identify the animals we encounter on walks. They take notes and make sketches of what they see along the way. Pack your bag with what you'll need to jot down notes and capture every detail you see. Remember to be prepared and carry anything you might need in case of an emergency—from a map to a first-aid kit!

2 CHOOSE A SPOT, HEAD OUTSIDE, AND OBSERVE

Keep your eyes open to spot the various animals living around you. Whether you're in your backyard, in the park, or in the woods, remember that the sound of your steps will alert animals to go into hiding. Tread lightly, and try standing still every once in a while. After several minutes of stillness, you may see or hear animals that were nowhere to be found when you first arrived.

I always remind my kids to keep their distance, be respectful, and be careful not to frighten the animals, disturb their work, or encroach on their turf (don't touch or handle nests, webs, or other places critters frequent).

WHAT TO PACK

First-aid kit

Compass

Camera

Sketchbook and pencil for writing notes and making drawings

Water

Field guide and map

Sunscreen

Binoculars

Rubber gloves

The closest we may ever get to some animals in our midst is through the footprints they leave behind. By paying attention to the markings on the ground, we can do a little detective work to identify which animals have crossed our path. Wet and muddy areas are perfect for finding prints, specifically near water sources such as the banks of a creek, pond, or river. If you aren't sure what to look for, seek out a field guide that includes drawings or photographs of animal tracks.

③ FIND A WAY TO HELP

You may have a favorite animal living near you. There are lots of options for helping your local wildlife. Consider planting, building, or crafting a special item on your property that will help an animal in need. You may choose to plant a pollinator garden to attract butterflies and hummingbirds, or build a bee house to help the mason bee lay her eggs. You could create or purchase a squirrel feeder to give squirrels some food of their own, make some simple bird feeders to keep the birds happy, or put window stickers on your windows to keep our feathered friends out of harm's way. You may choose to build a frog sanctuary, a nest for barn owls, or a simple device to keep your garbage cans closed so that raccoons, opossums, foxes, and other scavenging animals stay out of danger. You name it! There are plenty of ways to get involved and help.

④ HELP OTHERS MAKE A DIFFERENCE

Spread the word! Let others know about the nature walk you took and the items you crafted for your local wildlife. Share info about some simple tasks that help the animals, and maybe they'll be inspired to get involved and help, too.

IDENTIFYING FOOTPRINTS

Field mouse

Skunk

Squirrel

Opossum

Bird

Fox

Crow

Rabbit

Raccoon

Deer

HOW TO HELP AN INJURED ANIMAL

T**HE MOST IMPORTANT** information to keep in mind and communicate to others before going outside is what to do if you find an injured animal.

It's crucial that you first identify whether or not the animal is actually injured. Moving an uninjured animal can do more harm than good, especially if you are separating the animal from its family or taking it to an environment in which it can't survive. An animal does not need us to rescue it unless it is distressed or obviously injured. A baby animal's best chance for survival is to be with its mother. For example, baby birds of many species spend as many as two to five days on the ground before they can fly. This is a normal and vital part of the young birds' development. While they are on the ground, the birds are cared for and protected by their parents and are taught life skills such as finding food, recognizing predators, and learning to fly. If a baby bird is not injured, it is essential to leave it where it is so it can learn from its parents.

However, if you find an animal on the ground that is bleeding, has a broken limb, or is shivering and distressed, it is likely that this animal could use your help. Your next and most important step is to call your local wildlife rehabilitator to assess whether the animal truly needs human assistance. It is important to refrain from picking up or moving the injured animal before speaking to a professional, as you could cause more damage to the injured animal by moving it, no matter how careful and gentle you are. It is also possible that by moving the injured animal, you could

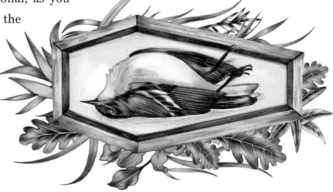

increase its fear, which might lead to the animal's biting, scratching, or otherwise attempting to hurt you.

Finally, there are a lot of animals in the animal kingdom that rely on their senses, in particular their sense of smell, to find out what is going on in their environment and to recognize their family and friends. By touching an injured animal, you are passing on a new scent to that creature, one that could potentially deter its own mother from rescuing or helping it. Although you may want to act instantly and do all you can for the injured creature, the best action you can take to ensure its best chance for survival is to call an expert and to stand guard while you wait for the expert's assistance.

WHAT IS A WILDLIFE REHABILITATOR?

A wildlife rehabilitator provides professional help to sick, injured, or abandoned animals found in parks, woods, and neighborhoods. In most states, it is against the law to keep wild animals as pets, so rehabilitators maintain special permits in order to diagnose the animals, administer first aid, apply physical therapy, or perform other necessary treatments to nurse the animal back to health and safely return it to the wild.

Before you go out on your Hurtless Hunt, be sure to research and note the contact number for your local wildlife rehabilitator, just in case you find an injured animal. You can visit the website of the National Wildlife Rehabilitators Association (NWRA) for links to licensed rehabilitators in specific states, or to find a contact number for a member of the NWRA in your area. Phone listings for your local wildlife rehabilitator may also be available through your town or county animal control office, veterinarian, or humane society. Alternatively, you could check the Yellow Pages or look online at websites such as Wildlife Rehabber.

STEWART FAMILY TRADITIONS:

The Great Backyard Bird Count

I HAVE ALWAYS MARVELED at my children's ability to talk so often and for so long. I, on the other hand, only manage to have something to say on average once every hour and can usually accomplish it in three sentences or less. Often it's "I'm hungry," followed by, "I have to go to the bathroom," followed by, "Now, why did I walk into the kitchen?"

Talkative as my kids are, however, there is a family tradition that silences them for the longest interval of the year, and that is the Great Backyard Bird Count, an annual event held by the Audubon Society each February. The event lasts four days, during which individual bird-watchers spend at least fifteen minutes a day counting birds in as many places as they can and making their best estimate of how many of each species they've seen. At the end of the event, participants submit their findings, and the results go toward helping researchers collect data and learn more about bird populations.

Armed with field guides and notebooks, we head out into the woods to be scientists. I love telling my kids that animals in the woods can hear us coming from miles away, and that they're not going to relax and start going about their usual business unless we can find a spot where we will be comfortable staying for fifteen minutes without making a peep. It works like a charm and is probably the closest we'll ever get to a family meditation. In shared silence, we delight in spying the brightly colored common yellowthroat warbler, the adorable Carolina chickadee, and the regal snowy egret.

MY NEW JERSEY GANG

Indigo bunting

Red-winged blackbird

Kestrel

Common titmouse

American redstart

Blackburnian warbler

Belted kingfisher

Little blue heron

Eastern meadowlark

Northern gannet

Black-bellied plover

Black-crowned night heron

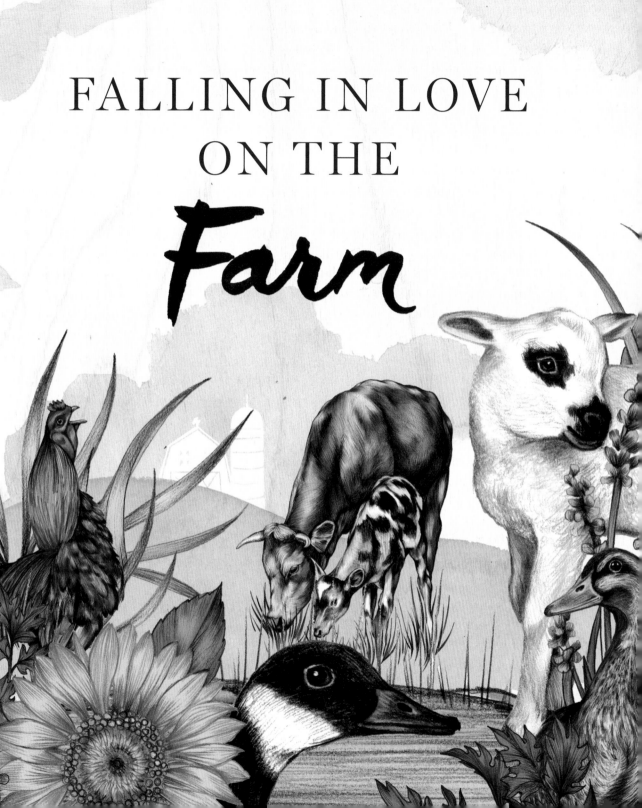

FALLING IN LOVE
ON THE
Farm

I**F YOU ARE A PARENT AND SOMEONE ASKS** you to describe the happiest day of your life, you'd best be prepared to respond with the day your child was born. If, however, you are indulged and asked what your next happiest day is, odds are you're going to give a much more interesting answer.

My next happiest day happened in August 1999 on a farm in Hudson Valley, New York. After years of working in design-related jobs, I realized it was time to follow my heart, so I headed back to school to get a degree in veterinary technology. Large-animal medicine was part of my studies, which is how I found myself spending a week at Robby's Dairy Farm, where my fellow students and I learned that there was a pregnant cow named Miss Eyebrows that was downed (that is, lying on the ground for more than twenty-four hours and not able to get up).

A downed cow can be down for many reasons: an injured leg, a pinched nerve, a calcium imbalance, or other kinds of sickness. It is a dangerous position for a cow to be in. As she lies on her side, her sheer size and weight can cause her to lose circulation in her legs, which can lead to an inability to stand up—and that can be fatal. Tragically, there is a tremendous amount of undercover video of these cows being brutalized in slaughterhouses and industrial farms.

Our professor, a large-animal veterinarian, took Robby aside after examining Miss Eyebrows and within earshot of all the students told him that the prospects for this cow getting up were not good. He went on to say that if Robby wanted, he would make a call to a nearby friend who would be able to get him a good price on the meat of the cow. We students were crestfallen. Then Robby spoke: "These girls give me my livelihood, and in return I give them a good life. They live, die, and are buried on this property." He then went on to describe how a farmer friend of his was able to get a downed cow back on her feet by carefully lifting her into a crane, placing her gently into a nearby pond, and massaging her muscles until her circulation returned. Off Robby went to get the crane.

Once the cow was in the water, our professor turned to us to ask who wanted to get into the pond and start massaging. Without hesitation, I ran into the water in my scrubs and rubber boots. I even chose the back end. Standing there in the murky pond water made murkier by Miss Eyebrows releasing several bowel movements from a combination of stress and, I hoped, relief, was almost my next happiest day, but the actual next happiest day was to come a few days later on August 6, when I arrived at the farm to find that Miss Eyebrows was standing and was about to give birth—and since it was my birthday, I was chosen to assist.

Days of struggle and exhilaration had produced a beautiful, healthy calf, born of a cow that on another person's farm might have been slaughtered for meat. Only after the calf was born did we notice that all of the other cows, which had been scattered far out in the pasture prior to the birth, were now lined up against the fence looking to get a view of the new baby that would be joining their family. I was hooked on cows and on the beauty of the relationship between animals and humans, and this was by far my next happiest day.

My Superheroes

WHILE WE MARVEL at exotic animals in the wild, we often forget just how close we are to true marvels of nature. The farm is home to inspiring mothers, acrobatic geniuses, intensely loyal friends, and incredible intelligences. They are superheroes hiding in plain sight. In the best and the worst of circumstances, they are there with emotional strength, physical power, telescopic vision, enhanced memory, extraordinary senses, and forbearance.

Our relationship with farm animals is perhaps the most fraught and complicated of all our relationships with animals. Our lives are very much entwined with those of farm animals, whether we live on a farm or not. Because the majority of us use farm animals in some way—whether by eating them or what they produce, or by wearing or otherwise using the attributes of the animal—we are in the position of both caring for farm animals while also harvesting them. The growth of the industrial farm has not only adulterated the foods we eat but also has horribly mishandled the animals in its care. Animals are treated as products and not as sentient creatures, and their demise usually comes with unnecessary pain and stress. We owe these noble, trusting creatures so much more. As Temple Grandin, advocate for autism and animal rights, so poignantly says, "Nature is cruel, but we don't have to be."

We all learned pretty early in life that the cow says, "Moo." What we probably didn't learn is why. If we start thinking about farm animals as sentient creatures, we may have to change the way we live. Human nature usually rails against this. In this chapter, I will ask you to be brave and keep reading, not because you'll learn things you don't want to know, but because you might fall in love—and we all know that falling in love can sometimes be a lot scarier.

THE COW:
Eternally Maternal

I WOULD JUMP IN front of a train to save my family, my dogs, and any cow. Cows are my favorite animal of all those that can't live in my house. Their large doe eyes, warm breath, and wet noses warm the cockles of my heart. My son says that the cow is his favorite animal, as well. When he was very young, he asked me why I didn't eat hamburgers. I said that I didn't eat hamburgers because I love cows too much. He said, "I love them, too! It's so nice how they give us hamburgers and milk." I know that at the time, he imagined that cows gave us these things in the same way that a soft-serve machine painlessly serves up delicious ice cream, but I understood what he meant. Back then my husband was very carnivorous. Due to our mixed marriage we have decided to present both diets to the kids and let them come to their own decision about how and what they want to eat when they're old enough to fully understand all the implications of their choices. And so I said to my son, "Yes, cows are very, very, very nice!"

Cows are deeply emotional animals. They are remarkable mothers, exceedingly sensitive, and attuned to their friends and family. They have a strong need for social interaction. They love to be together. They pay attention to others' behavior and act accordingly. They love soothing music. They cry. They never forget if someone has wronged them. To me, that sounds like the perfect friend for those who strive to be good friends themselves.

Farm Sanctuary's rescued calf, Nik, and the mom he wished he'd had

NURSING

Like all mammals, cows produce milk to feed their young. When a cow becomes pregnant, she churns her food into milk, which is created and stored in the udder, ready for the calf to feed on once it is born. Some cows will nurse their calves for up to three years.

BONDING

The bond between a mother and her calf is unbreakable. A mother cow carries her baby inside her stomach for close to 290 days, just like a pregnant woman. Long after a female calf is born and matures into an adult cow, she and her mom remain side by side, and they graze together for years. If separated from her calf, a cow will walk miles to find her.

BABYSITTING

Cows arrange to babysit for one another. One cow will care for all of the calves in the herd while the other adults head out to graze. The babysitter stays with the calves and watches over them.

A GENTLE APPROACH:

How to Interact with a Cow

EVERY TIME I meet a new animal, I secretly hope that the animal will feel an immediate kinship to me. Along with my childhood dream of being Charlotte the Spider, I also wanted to be Dr. Doolittle. To stack the deck in my favor, I research how to approach most animals before I meet them. Below are a few things you need to consider before approaching and as you interact with a cow. Remember that all cows are different, and their moods can vary from day to day. Never approach a cow that is not accustomed to your presence, and be sure to have an expert around at all times.

1. **Stay in their field of vision.** Cows have almost 360-degree panoramic vision, but like your car, they have a blind spot directly behind their rear ends. They're not so great at telling how close or far away an object is, either. So don't sneak up in a cow's blind spot or you risk frightening her and causing her to flee.

2. **Be aware of their strong tongues.** Cows' tongues are like glorious exfoliators. Cows have a limited number of teeth, so their tongues are extremely strong. They pull up grass, swish it around, and pass it to the back of their mouths where there are more teeth for chewing. If a cow sticks out her tongue to lick your hands, go ahead and enjoy the exfoliation, but don't let your fingers get too far to the back of her tongue or mouth.

3. **Keep clear of swinging heads and tails.** Cows are not usually aggressive, but man, are their tails strong! It's natural for them to be swinging their heads and tails from side to side to brush off or swat at flies. Don't get caught in the crosshairs—you could get a nasty (though totally unintentional) bump.

4. **Be sensitive.** Cows are gentle and appreciate being petted on their bodies and legs. Gently running your hands over their bodies and brushing their coats can relax them and allow them to become accustomed to your touch. A light scratch under the chin or behind the ears will go over nicely. If a cow shies away from your touch, don't force it. Try another time!

5. **Pay attention to who else is around.** Cows are very protective of their herd friends and calves. If a cow sends out signs of aggravation—growling or head tossing—it's best to get out of the way. A protective cow mom or friend will charge if feeling threatened, and you don't want to be caught in the middle.

Handsome Mario, from Farm Sanctuary California

WHAT MAKES A COW UNHAPPY?

MOOOOSIC IS NOT always music to a cow's ears! Cows moo if they are feeling distressed or in pain, panicked, or lonely. They moo when they are hungry (so do I sometimes), separated from their friends, or in heat. But the most sorrowful-sounding moo of all can be heard if a cow is separated from her calf. While staying at a friend's farm I was rocked to my core by hearing heifers wail for four days straight. A neighboring farmer had separated six calves from their mothers. He kept the cows on his property to take advantage of a farm assessment and then sold their calves for veal. You could hear the mamas' cries for miles.

This type of separation happens around the clock at industrial farms. Cows in these factories are artificially inseminated year after year so they will lactate and produce milk. This was not something I was aware of during my coursework at Robby's Farm. As devoted mothers, their hearts break each time their babies are taken from them, with the calves shipped off in individual crates. The female calves are moved to small groups where they are fed a milk replacement and pass time as they wait to join the dairy line like their mothers. This is sadly only one of too many indignities they suffer.

Sometimes you have to listen a little harder to understand what an animal is trying to say. And sometimes they are saying it so loudly it's hard to imagine people don't hear.

Make

NATURAL FLY DETERRENT

Many of the cows at Farm Sanctuary, having come from industrial farms, have had up to two-thirds of their tails docked. Just like the docking of a dog's tail, docking a cow's tail is an incredibly painful and often unnecessary procedure.

Cows use their tails to prevent pesky flies from biting their skin. These flies and their bites have been shown to cause disruption to a cow's routine and growth, so docking truly puts them at a disadvantage when they are simply trying to fight back and avoid the bite! A more humane and appropriate alternative is the use of natural fly deterrent. Let's give our bovine buddies an extra hand and choose a more humane method of fly control.

SUPPLIES

- **1 cup vinegar**
- **1 cup water**
- **Essential oils, such as lavender, citronella, eucalyptus, and pennyroyal**

DIRECTIONS

1 Mix the vinegar and water together in a bowl.

2 Add drops of your favorite essential oil.

3 Pour into a spray bottle.

4 Spray on surfaces that flies frequent, including a cow's hide. Be sure to avoid spraying on the face, eyes, ears, mouth, or genital area.

THE PIG:
The Brainiac of the Farm

WILBUR WAS THE first pig I ever fell in love with. It was 1973, and the very first movie I saw in the theater was *Charlotte's Web*, based on the famous book of the same title. My mom was already well aware of my affection for animals and thought there was no better starter film for me. She was right. At the tender age of six, I vowed to someday be a Charlotte myself. I would go on to fall in love with every pig I ever met. My porcine pals shared the qualities I appreciated in the pig made famous in the story and film. They were all sensitive, intelligent, and longing for companionship . . . and, boy oh boy, were they cute! If you've never snuggled with a pig, you haven't lived. A friend of mine who suffers from migraines even tells me that her headaches are significantly relieved when she lies with her pigs.

It's a cruel twist of fate that pigs have not shared the same good fortune that many of our companion animals have today. Aside from their rough coats and high-pitched protestations, they are essentially a smarter version of our beloved dogs.

I try to always provide the best environment for my pigs. This consists of a big mud puddle, a clean pond, shelter for shade during the summer and warmth during the winter, and a roomy area for building communal nests. Female pigs burrow and build a nest for their group of fellow pigs and piglets to sleep in together and cuddle up nose to nose. They know how to make the home cozy for their piglets and are very good at keeping the space clean and sanitary for their roommates, as they hate mess!

My pigs, Pugsley and Christopher

WHAT MAKES A PIG UNHAPPY?

AN UNHAPPY PIG is one that's forced to live the majority of her life in a gestational crate. A gestational crate is a metal cage approximately 2 feet wide that sits on a concrete floor and closely confines a female pig, or sow, preventing her from moving or turning. Its purpose (and the pig's, too) is simply for repeat impregnation. The sow is kept in the crate for the duration of her pregnancy. Once she has given birth to her piglets, she nurses them and is then impregnated once more. In intensive pig farms, a sow will have an average of 2½ litters every year for three or four years, after which she is slaughtered.

Without the ability to explore, build nests, forage and root, or form social relationships, pigs are not only bound to be unhappy, but there are those who say it causes them to go insane. Physical and mental injuries set in quickly. They develop crippling leg disorders and unhealthy weight gain. They become prone to painful bone and muscle injuries, including bleeding jaws from chewing on the bars in an attempt to break free. Pigs in crates run the risk of suffering respiratory disorders from being exposed to high levels of ammonia from their own feces.

THE PERFECT PIG PALACE

J UST LIKE CATS and dogs, pigs need environmental enrichment to help them thrive and keep them from becoming bored, frustrated, or fanatical. They are highly intelligent animals. There are many ways of providing enrichment to pigs to engage their natural behaviors and excite their intelligence.

❶ ROOT VEGGIE GARDEN

Rather than simply supplying all of their food in one place, let pigs forage. They are natural foragers and love to dig up different root vegetables such as carrots, turnips, and beets. Access to root vegetables from the garden allows pigs to engage their rooting ways and use their smarts and senses to find their food. Try burying some root vegetables in your pigs' pen and letting them dig for them using their snouts.

❷ CHEW TOYS

Most pigs love toys. But unlike dogs, they prefer their toy be kept clean; otherwise they lose interest. Pigs have to throw things to the back of their mouths to chew. Their front teeth act as scoopers. They are at risk of choking on items like whole potatoes or tennis balls. I make sure my pigs' toys are at least the size of jolly balls or bowling balls.

❸ STRAW

Straw is easy for pigs to manipulate and root in. It can be used as bedding or can be mixed with feed to allow pigs to dig for their food. Burying food in straw gives them the opportunity to play and forage to their hearts' content.

Farm Sanctuary's
Sprinkles

THE REAL PIG LATIN:

Learn to Speak a Pig's Language, as Modeled by Farm Sanctuary's Ramona

I REMEMBER HEARING ABOUT pig latin as a youngster and initially assuming it was the actual language of pigs. Sadly, even today many animal researchers and behaviorists are disappointed by how little research there is on this topic. It's difficult to get funding to study the behavior and feelings of animals when people worry that it might impede their ability to enjoy a BLT . . . I'm talking to you, husband!

Pigs most certainly have their own language. They are smart, funny, respectful, and complex. They have a wide range of movements and sounds that they use to communicate with one another.

"LET'S PLAY!"

A pig that runs quickly toward a human or another pig, swaying his head and hips back and forth and spinning in a circle, is letting you know he's ready to play.

"TEE-HEE-HEE!"

Pigs have a sense of humor! A nervous pig will let out a little bark if startled. The sound can make people run for miles, and pigs find this very funny. They often use this sound to initiate play.

"IT'S NAP TIME."

A relaxed and tired pig will lie down in the pen with his belly exposed. And, yes, if you're a friend, he wouldn't mind if you rubbed his belly.

"I'M SAD."

You can tell if a pig is sad or depressed simply by looking into his eyes. He becomes lethargic, lies down on his belly, and keeps his head low and lazy.

"I'M THE BOSS."

When two pigs meet and one makes a growling sound, the growly one is showing his or her rank in the pecking order. The lower-ranking pig will often make a grumbling sound but will usually move out of the way as asked. Pigs appreciate a peaceful home. The head of the tribe can simply walk through the others and take any bed she likes, drink from whichever water unit she chooses, and sit with whomever she pleases. The other pigs might object with small grumbles, but rarely do they do anything besides quickly move out of the way once a hierarchy has been established.

"I CHOOSE YOU!"

Pigs will choose their own very best friend. They pick one pal to spend time with, sleep with, forage with, and play with.

THE GOAT:

A Kid at Play

Farm Sanctuary's
mother-daughter rescue,
Ginger and Curry

IF YOU'RE EVER feeling down in the dumps, get thee to a goat pasture. Goats are probably the most entertaining creatures to watch on the farm. They are like forever-curious toddlers looking for fun. A baby goat is aptly called a kid. Every object strikes goats as potential playground equipment. Put a goat behind a six-foot fence and he will break it open with his horns in the name of freedom. Place a small object in her way and she will calculate the different angles to get past it. Fill a field with trees, and goats will climb the branches to reach the highest level.

Female goats are nurturers. They are patient, maternal, and often even foster other young animals, including lambs, calves, and foals. They are curious and friendly creatures constantly looking to have a little fun. When I look into goats' animated faces, I swear I can feel the endorphins release.

Farm Sanctuary's Peanut

WHAT MAKES A GOAT UNHAPPY?

MOST PEOPLE OFFHANDEDLY suggest that goats and sheep are the two animals on the farm that have the most in common with each other. But very few of us stop to think about the deep emotional similarities that goats share with cows. Female goats, just like female cows, are known for being incredibly patient, loving, and nurturing moms—and like cows, they all too often fall victim to cruel separation from their babies. Large-scale farms will impregnate their female goats in order to get them lactating and collect their milk. After the goats give birth, the kids are taken away from their mothers and the mothers are forced to reenter the dairy line. The female kids are raised as replacements for dairy production, while the males are either slaughtered or sold for meat. To eliminate "buck odor," males are cruelly castrated with little to no anesthetic, and many male and female goats are painfully disbudded (dehorned).

Just like a cow, a goat produces milk for the consumption and health of her babies and will churn just enough to satisfy their needs.

MAKE YOUR GOATS' PLAYGROUND SAFE

Within every goat there beats the heart of an intrepid explorer who thirsts for new adventures and uncharted territory. When it comes to fencing in a goat, you need to remember that you are going up against a crafty escape artist. Goats are inquisitive and intelligent and see no harm in breaking down barriers. They'll use their horns and acrobatic ability, and in some instances I suspect computer-rendered blueprints have been involved.

When goats get out of their enclosure, besides worrying about them falling into the clutches of predators, we worry about these curious browsers getting within reach of plants that are toxic. The list of those that are harmful to goats is long. Rhododendron, azalea, and mountain laurel are just a few that can cause poisoning and interfere with skeletal muscle, cardiac muscle, and nerve function. It's important to pay close attention to the wild shrubs, plants, and grasses that grow in or near a goat's home and get to those that are dangerous before the goats do!

1. Goats are skilled climbers and enjoy testing their coordination. Climbing equipment encourages exercise and improves a goat's health.

2. Goats have amazing balance and like the challenge of walking along items of different heights and widths. They can negotiate a width as thin as a tightrope.

3. Goats jump over and crawl under any object they deem fun and interesting. Tires are good for this, and filling them with cement can help to trim goats' hooves.

4. Goats can spend hours amused by a seesaw and the effects of weight shift on each side. Goats entertain themselves for long periods by playing together in pairs or in groups.

FARM SANCTUARY'S KIDS LIVING THE GOOD LIFE

THE SHEEP:

The Sweethearts of the Farm

SHEEP DON'T HAVE a mean bone in their body. They are exceedingly observant and responsive to other sheep, animals, and humans alike. In fact, they have a specialized part of the brain that enables them to recognize and remember up to fifty different faces. Sheep gravitate toward calm and smiling faces and avoid angry or anxious ones even when those are associated with food. Deep within that soft, woolly exterior lurks a creature with a rich social and emotional life. So, the next time we describe someone as sheepish, we may want to stop and think if in fact we mean to be saying that this person is a socially aware, emotionally complex creature who likes the comfort of a cuddle. I strive to be sheepish.

Farm Sanctuary's adorable duo, Olive and Aiden

WHAT MAKES A SHEEP UNHAPPY?

NORMALLY, SHEEP GROW just enough wool to maintain their body temperature in both the warmer and colder months, but through selective breeding, sheep are made to produce unnaturally high amounts of wool for shearing. If a sheep's living conditions cause the animal to overheat, then shearing may be in his best interest. As creatures that scare easily, sheep can find the shearing process stressful and traumatic when it's not done correctly. The heat from the shears scalds their skin, and the shears can hit them with great force as a result of rough handling and the animals' struggles to break free. Sheep can be badly cut from shearing that is done by those in need of speed.

THE SHEEP'S DREAM GARDEN

DON'T ASK WHAT possessed me, but last spring I purchased a life-size resin sheep for my front yard. I was truly embracing life in New Jersey. Anytime an animal, real or resin, comes into our home, my daughter is quick to start her research. She was intrigued by sheep's medicinal use of plants. When grazing in pastures, sheep are usually able to learn what they can and cannot eat, choosing different plants to help fight diseases and learning about the benefits of each plant as they eat it. Fascinated by what she had learned, we decided to plant a medicinal garden around our new sculptural friend. I like to believe that this makes my purchase a little less strange. Right?

1 **Lavender.** The entire lavender plant is useful for sheep as an antiseptic, antifungal, and antibacterial.

2 **Marigold.** Marigold is a good heart medicine that sheep (and goats) love to graze and browse on.

3 **Dandelion.** An overall good health conditioner for sheep, dandelion can be beneficial for blood cleansing, teeth strengthening, and curing jaundice.

4 **Parsley.** Sheep love to graze on parsley as it is rich in iron and copper and can improve blood flow. It also contains vitamins A and B, which are good for diseases of the urinary tract, arthritis, and rheumatism.

5 **Raspberry.** Raspberry aids in digestion, and sheep love it as a choice in the fields.

6 **Sunflowers.** The stalks and heads of sunflowers can be helpful in a sheep's diet. Although they are low in protein, they are high in fiber and are rich in vitamins A, B, D, and E.

7 **Thyme.** The oils in thyme leaves are beneficial for repelling worms in sheep.

8 **Watercress.** Watercress is high in vitamins A, B and C, as well as iron, copper, magnesium, and calcium.

COUNTING SHEEP:
Personalities Abound!

I'LL ADMIT I'VE tried to fall asleep using the technique of counting sheep jumping over a fence, but rather than producing the intended result, it just left me excitedly imagining having a flock of my own. Take a look at my adopted flock and I think you'll see why even science shows that this technique doesn't work very well. These sheep are all one-of-a-kind wonders!

LOVERBOY CASH

Cash is a lover. He's the first sheep to greet guests when they visit and can't get close enough when they pet him. Cash doesn't understand why he can't just sit in your lap. He's downright dreamy.

MAMA'S BOY DANIEL

Daniel can always be found at his mother's side. His mom arrived at the sanctuary pregnant and gave birth to Daniel, a cute, wrinkly little red-headed lamb who was full of personality just hours after he was born.

BEAR THE SHEEPLE

Bear is a people sheep—or a "sheeple." She loves her caregivers, who raised her from the time she was just a few hours old, and is always very busy with some sort of project. She is obsessed with her mineral block and loves her best pal, Carlee, who is just half her size.

MADELINE THE SWEETIE

Madeline is sweet, loving, and independent. She was raised with a goat, a sheep named Clarabelle, and her sister Samantha. Samantha almost died when she first arrived, and Madeline spent the first month in the hospital with her. She prefers the company of other sheep, but there are a few special humans that she's let into her flock.

KATHERINE THE BOLD

Katherine is a comedian who is known for drinking out of the drains. When she hears a drop, she runs out to capture the cold water. Her best friend is her twin brother, Will, whom she loves to spend time with. Together they love to sleep together with their mom, Elle.

RAMSEY THE GREGARIOUS

Ramsey is a Scottish Blackface sheep. He came to Farm Sanctuary with his cousins, his mother, and his aunts. He is part of a very tight and protective family. The mothers let the babies ride on their backs as the other family members circle around the young to protect them.

OTHER-ABLED GRACE

Grace recently went blind and relies solely on sound to find her favorite sheep and people. She loves being scratched on the chest, and if you stop too soon, she paws you to get more attention. She is loving and kind.

SHY JOANNE

Joanne is gentle and a complete opposite to her twin brother, Brian, who is quite an extrovert. The twins were rescued from a lamb facility, where moms had their babies taken from them at a very young age. The two siblings still loyally sleep at each other's side at night.

FUN-LOVING FRANCIS

Francis is a fun-loving sheep who adores head-butting and playing with the others in his flock, as well as humans. He spends a great deal of time with guests and loves to rub his head all over their jeans. He could stand for hours getting petted.

THE HORSE:
The Intuitive Soul

MY RELATIONSHIP WITH my horse, Alegría, is like none I have with any other living creature. As I look into her warm, kind eyes, I am immediately calmed and humbled. When I return home after spending time with her, I can become frustrated with others for their inability to read my mind as well as she seems to. She is extremely in tune with me. When I shift my weight back in the saddle, she stops. When I look to the right, she heads right. And with just the subtlest pressure of my right calf, she is nudged left. I always make sure she is enjoying herself on our rides as much as I am, and that our adventures satisfy both our curiosities. After all, she is gracious enough to tolerate my weight on her back. It is never lost on me that I am sharing a walk through nature with a member of another species. The word *alegría* means happiness, and that is certainly what she gives me.

My horse, Alegría

WHAT MAKES A HORSE UNHAPPY? CITY TRAFFIC

IN CITIES ACROSS the United States, draft horses pull carriages in heavy traffic. Having experienced the stress of New York City's streets myself, my heart goes out to these majestic animals navigating thoroughfares jammed with impatient cabdrivers, loud buses, and never-ending construction. Sadly, accidents have occurred. Spooked by all the commotion, horses have bolted through traffic and have accidentally hit vehicles and vice versa. Horses and people have been injured, and some horses have even lost their lives.

The tradition of horse-drawn carriages seems to be becoming less and less safe as our cities become more chaotic and crowded. In these days of John Deere tractors, electric cars, and creative problem solvers capable of coming up with ingenious and efficient ways to work and live, do we still need to use and breed animals as beasts of burden? Like the animals in circuses and in many zoos, these animals have become the means to an end for human profit. It is possible to find joy in and inspiration from animals without their having to suffer for it.

HORSE-ESE:

Learn to Speak a Horse's Language,
As Modeled by My Daughter's Horse, Steady

WHEN WE DO our homework, are patient, and spend a lot of time just observing our equine companions, we can reap the benefits of the very special relationship to be had between human and horse. With just the swish of a tail and flick of the ears, a horse will let us know how she is feeling and give us the chance to respond and connect. Horses provide immediate feedback to us about what we are communicating with our body language and movements, which allows us the opportunity to establish trust and feel more comfortable with them. Horses set a wonderful example of being patient and intuitive when it comes to understanding our desires. Since ancient times, horses have been believed to have an affinity for healing youth and an ability to reduce anxiety. There's a general consensus today that working with a therapy animal can be a highly beneficial addition to treatment programs for children with autism or Asperger's syndrome.

LISTENING

When you're riding, a horse with ears swiveled back is saying, "I'm really concentrating on and listening to you."

IRRITATED

An annoyed, sick, or nervous horse will swish her tail (not at flies), lift or stomp one hind leg, and warn you that she needs to get out of here.

FEARFUL

A frightened horse will swing
her hindquarters toward you
and look as though she is
ready to kick.

ANGRY

A horse that is ready to
buck, kick, or bite will pin
her ears back flat against
her head and fiercely swish
her tail. This gal is warning
you to stay away!

FRIENDLY

A horse that wants to say hi tilts her
ears forward and stretches her neck
and head toward you.

CURIOUS

The horse with ears pricked
forward and head held high is
wondering, "What's over there?"

A GENTLE APPROACH:

How to Interact with a Horse

T**HE WAY THAT** we approach a horse can set the tone for current and future interactions. Here are some tips to consider before and during any interaction with a horse.

1. **Be aware of blind spots.** Horses have two blind spots: directly in front of their noses, and directly behind their hindquarters. Approaching a horse from either of those directions, and especially from behind, is a bad idea. Although horses have large eyes and the ability to see for miles in nearly all directions, they are still easily startled.

2. **Speak softly and gently.** The best way to approach a horse is to gently call out, ensuring that she sees you; slowly come near; keep your body language calm and quiet; and whenever possible, approach from an angle rather than directly in front or from directly behind.

3. **Stay on the horse's left.** Horses are customarily trained to be approached, led, worked with, bridled and saddled, and mounted from the horse's left, so that is the best side from which to approach.

4. **Offer a treat.** Holding your hands out to allow a horse to sniff your fingers is a smart way to help her get accustomed to your scent and presence. Bringing a treat will also help to get on her good side. Place it in the palm of your hand, fingers flat to avoid accidental nipping, and allow her to retrieve is when she is ready. I usually keep my palm flat against my horse Alegría's lips until I know she's gotten a good hold on the treat. Pet her gently on her face, neck, and sides.

My daughter's favorite senior pony, Rocky

Bake

HOMEMADE HORSE COOKIES

I like to believe that our horses, Alegría and Steady, adore our family because of our gentle natures and great handling, but I sometimes wonder if it's because of the delicious treats my daughter and I make for them.

   This recipe makes approximately 18 cookies.

INGREDIENTS

- **4 cups hulled or rolled oats**
- **1 cup flour**
- **½ cup molasses**
- **¾ cup water**

DIRECTIONS

1   Preheat the oven to 300°F and grease a cookie sheet.

2   Combine the oats, flour, molasses, and water in a large mixing bowl and stir until the mixture reaches the consistency of a thick dough.

3   Place tablespoons of dough 1 inch apart on the cookie sheet.

4   Bake the cookies for 30 minutes or until crisp.

5   Allow them to cool for approximately 30 minutes before feeding them to horses.

AN ODE TO THE BIRDS ON THE FARM:

Chicken, Turkey, Duck, Goose

MANY FOLKS FIND it much easier to relate to mammals than to birds—which can make life unfortunate for chickens, turkeys, ducks, and geese, who go largely unrecognized for their intelligence, emotions, and need for affection and attention.

I have complete admiration and compassion for chickens. They are incredibly brave and devoted to their offspring. A rooster will sacrifice himself to a hawk rather than allow the hawk to attack one of his hens. A mother hen's maternal instincts guide her every move. Chicks that have been separated from their mothers will forgo food rather than be separated from their siblings. I've had the pleasure of seeing a chicken melt trustingly into my arms as I've massaged her neck, even though she previously had been horribly abused.

Similarly, turkeys are immensely affectionate. They love to be caressed and even make purring sounds when they are content.

Ducks are very social and resourceful. A mother duck (also known as a hen) will lead her ducklings as far as a half mile or more on land after they hatch to find a suitable water source for swimming and feeding.

Geese are indeed birds of a feather that flock together. If a female goose's mate or gosling becomes sick or injured, she will often refuse to leave his side even at the risk of her own survival. Who needs the excitement of a UFO sighting when you can just look for a gaggle of geese if you want to observe intelligent life forms flying overhead?

MEET THE FLOCKERS:
Family First

EACH MEMBER OF the Flocker family has his or her own story to tell and a significant role to play. The family is made up of Mr. Flocker (the rooster), Mrs. Flocker (the hen), and the Flocker siblings—a couple of chicks. This feathered family has among the most intense relationships on the farm, displaying extreme protective and connected qualities. Let's take a look at each and the roles they play.

FLOCKER SIBLINGS

Chicks are highly attuned to their mothers and to one another. The Flocker siblings look up to their mom as teacher and protector to whom they owe everything. Chicks will climb under their mother's wing to stay warm and safe while they observe their surroundings. They can slide right underneath a hen's body if they feel they need to be protected, knowing that the mother hen will always keep them safe.

MRS. FLOCKER

Not for nothing do we call hovering maternal types "mother hens." Hens are among the most devoted moms on the farm. Mrs. Flocker is caring and protective of her babies, both in the shell and once they hatch. A hen will encourage chicks to search for food by teaching them how to scratch. She will also teach her chicks to recognize her call so that they come when she has found food for them to eat.

MR. FLOCKER

Roosters "rule the roost." Mr. Flocker's tough and dominant nature allows others to feel safe and has him sitting at the top of the pecking order. The rooster will sacrifice his own life in order to protect a hen, chick, or friend. He will stand in the way of a predator and risk being taken rather than lose his friend or family.

THE COMFORTABLE COOP:

The Perfect Family Home

AS A MOTHER myself, I appreciate how hard a hen works to keep her brood happy. It is very stressful to a hen when she feels her family is not safe. Giving her a well-protected, well-built home is the least we can do for her and will help to keep this mama calm and contented. Here are some features of a henhouse worth crowing about.

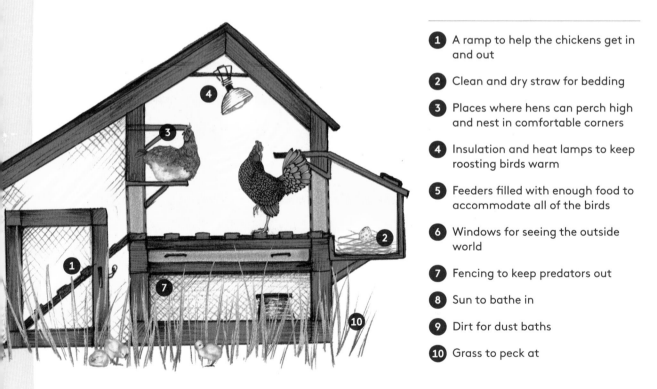

1. A ramp to help the chickens get in and out

2. Clean and dry straw for bedding

3. Places where hens can perch high and nest in comfortable corners

4. Insulation and heat lamps to keep roosting birds warm

5. Feeders filled with enough food to accommodate all of the birds

6. Windows for seeing the outside world

7. Fencing to keep predators out

8. Sun to bathe in

9. Dirt for dust baths

10. Grass to peck at

WHAT MAKES A CHICKEN UNHAPPY?

C HICKENS DON'T ASK for much. They just want a safe, clean home to live in as a family. Without sufficient space for carrying out their natural habits and behaviors, chickens can be plagued by many disabling and dangerous issues. A shed that is not large enough to comfortably house the chickens can lead to overcrowding, which can cause poor walking ability, thigh sores and scabs, scratches from other birds, lesions and rashes on the backs of their legs and feet, and the development of deep ulcers. A crammed shed can lead to a deterioration in air quality and higher rates of respiratory illness. Chickens living among urine- and manure-soaked materials can be exposed to excessive ammonia levels and are at risk for skin burns, ulcers, and painful respiratory problems. The fumes from excretory ammonia can be so strong that chickens develop a blinding eye disease so painful that the birds may rub their eyes with their wings and cry helplessly. Limited room to sleep can contribute to anxiety and nervousness, which stress the birds' hearts and lungs, and an insufficient environment in general can lead to fear, neurotic behaviors, and stress-related issues.

THE TURKEY:

An Underestimated Friend

I HIGHLY RECOMMEND TAKING some time to meet a turkey. Rub his warm, soft head and scratch his belly, and you'll most certainly feel conflicted at Thanksgiving. I've noticed that the turkey is one of the more misunderstood animals on the farm. People tend to assume that turkeys are just big, dumb birds without much personality or interest in interacting with people. Speaking from experience, I can assure you that that is just not the case! Yes, all turkeys are different, and no, not all turkeys like to be caressed, but there are ways of making a turkey happy. It's time to change the general opinion of these beautiful birds and see just how easy it is to make a turkey purr!

1. **Play some classical hits.** Turkeys will often cluck and gobble to the tune of certain songs, almost as if they're singing along.

2. **Admire their fans.** Turkeys have great big feathered fans, and male turkeys love to show them off. If a turkey is puffing out his feathers, pay attention and show some admiration.

3. **Offer some affection.** Only once you have become a frequent and trusted visitor should you attempt to pet a turkey. Turkeys love a good snuggle, stroke, or pet from a trusted friend. They can sit for hours nestled by your side receiving gentle pats.

4. **Allow families privacy.** Turkeys eat two main meals a day and really enjoy dining as a family. Mothers can be protective of their young and may attack, so when the family is gathering, be sure to stand back and give them privacy.

*Hank from
Farm Sanctuary*

WHAT MAKES A
TURKEY UNHAPPY

(Besides Thanksgiving)

I'**LL SPARE YOU** the details, but suffice it to say that these are just a few of the things that make turkeys unhappy: mutilations, confinement, scratches, ammonia burns from living in their own urine, stress-related aggression, de-snooding, de-toeing, and de-beaking. And that's just for starters. The treatment of birds in industrial farms is appalling! (Warning: Don't invite me to Thanksgiving dinner. I'm a real downer.)

MEET TOM, THE FLIRTY TURKEY

TOM, THE MALE turkey, uses a number of different features on his head and body in order to attract a female turkey, known as a hen. He is polygamous and is happy to mate with as many hens as he can attract. Luckily for him, his spectacular and unique facial features are sure to result in success with the ladies.

1 **The wattle.** This fleshy flap of skin hangs from the male turkey's neck and throat. The wattle turns bright red when the turkey is angry or during courting. Large wattles are a sign of good nutrition and high testosterone.

2 **The snood.** The snood is a fleshy mass hanging low over the beak and extending from the forehead. The snood will lay low when the turkey is relaxed, but when he is strutting it will engorge with blood and become elongated.

3 **The beard.** This group of hairlike feathers grows from the center of the breast. Some female hens also grow beards.

4 **The plumage.** Bright, colorful feathers spread across the male turkey's rear and make up a tail fan. The feathers are unique, featuring areas of green, copper, bronze, purple, red, and gold.

RATE YOUR SUCCESS

The male turkey has a mood ring head. During breeding season, a male turkey, known as a Tom or a gobbler, can express his feelings by the color of his head. Other turkeys—and farmers, too—can tell immediately from his head color whether or not it's a good idea to approach him. Use this as your guide.

- Red: If a turkey is feeling angry or aggressive, his head will turn wine-red. He will also lift his head, stretch his neck as far as he can, and make a trilling sound. Females do the same when they're annoyed, but without the red head.
- Blue: If a turkey is feeling excited or happy, his head will turn blue.
- White: If a turkey is feeling neutral, his head will turn white.

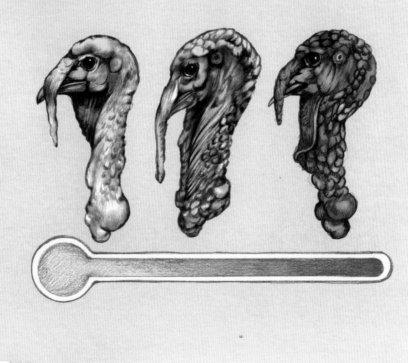

DUCKS AND GEESE:
The Sentimental Travelers

GEESE ARE LOYAL and protective of their partners and friends and will delay their migration if a fellow goose is injured and in need of help. They mate for life and mourn when their friends or family members die. Very few ducks actually "quack." Most whistle and growl, with females making the most noise. Misconceptions abound.

Ducks and geese living in public ponds and parks are put at risk by well-meaning grown-ups and children who enjoy feeding them stale bread and crackers. While this is a seemingly generous way to spend a day, it is actually causing more harm than good. Bread is not nutritious for the animals and fills their stomachs with empty nutrients. In addition, taking bread from human visitors becomes an easy option for receiving food, and so the birds lose their ability and desire to forage naturally, instead simply relying on handouts. Too much bread falling into the water and rotting there can cause pollution in the water, which can be hazardous to fish and other living creatures in the pond. Finally, feeding the ducks during the spring and summer sets them up for false expectations and disappointment in the winter months, when they need more food to stay warm, visitors are scarce, and the food supply that they have become accustomed to and dependent on is no longer as readily available.

Farm Sanctuary's Cinny the goose and Yerba the duck

WHAT MAKES GEESE AND DUCKS UNHAPPY? DOWN COMFORTERS!

DOWN FEATHERS FOUND on ducks and geese are valued for clothing insulation, comforters, and pillows. These fine feathers can be found close to the bird's skin, beneath their layer of external feathers. Ducks and geese have their throats slit and are then dumped in tanks of scalding water for the removal of large feathers, often when they are still conscious. Other times, they are simply dumped into the tanks without benefit of having their throats cut, to drown and scald to death. Some animals are live plucked, which means that their feathers are cruelly torn from their bodies while they are still alive.

Down feathers, however, are plucked from birds after they are slaughtered for meat or foie gras—although in some cases the birds are forcibly restrained and plucked while they are still alive. This method is more profitable because the birds can be exploited for multiple feather harvests. In both situations, the birds involved generally live short, miserable lives and die painful deaths for purposes that are absolutely unnecessary in this day and age, when there are so many thermal alternatives that are far superior to down feathers—including PrimaLoft, Thermal R, Omni-Heat, and Cocona insulation, to name just a few.

Giving Sanctuary

IN THE SUMMER of 2008, I headed to a rented place on the Jersey Shore for some rest and relaxation. Unfortunately, I forgot to pack a book for the beach. The previous renter had left a book on the coffee table entitled *Farm Sanctuary* by Gene Baur. The subtitle was *Changing Hearts and Minds About Animals and Food*.

This was trouble! My experience with the cows at Robby's Farm had already left me unable to eat once-delicious burgers. I knew that if I read the book, there'd be more changes to make, and I was scared. To complicate matters, I was in a relationship with a very carnivorous man. Would my becoming a vegetarian drive a wedge between us? What was I going to do when on our at-least-twice-a-week romantic outings my man wanted to go to Burgers, Burgers & More Burgers Bistro? I was not going to read that book.

But the cover photo of the author sitting in a beautiful pasture with a gorgeous cow by his side wore down my resistance. They really looked to be having a moment. The inevitable happened. I read the book.

Gene Baur founded Farm Sanctuary in 1986 after rescuing a live sheep from a pile of dead animals in a stockyard. His organization rescues discarded living animals from stockyards, slaughterhouses, and factory farms. At the sanctuary, the animals are provided with care and shelter for the remainder of their lives. The animals, staff, and supporters of Farm Sanctuary all serve as advocates for the humane treatment of animals, especially as it relates to modern-day industrial farming.

The things I feared would happen if I read the book didn't happen. Although I did become a vegetarian, I didn't mourn eating meat. Being more aware didn't cause a rift in my relationship, either. Instead it inspired both of us to do better. Following my conscience was so much easier than I had anticipated. Doing so brought many other positive changes in my life that I could never have anticipated, including the privilege of being able to call Gene Baur a friend. Thanks to him and to Farm Sanctuary's national director, Susie Coston, I've been able to adopt all the farm animals I ever dreamed of having.

IT'S TIME TO MAKE A CHANGE:

Let's Get Rid of Factory Farms

I BELIEVE ONE OF the most important things I can teach my kids is how to live healthy, compassionate lives. While the world offers many challenges to these lessons, none seem as ominous and diabolical as the impact that our corporate-controlled, industrial food systems have on our health, our children, and the animals we claim to care about. We attempt to carefully mete out antibiotics to our children while corporations give them to animals with abandon, raising animals so unhealthy that without antibiotics, they'd never make it to our tables. We cheer for the police officer who stops traffic to let a duck family cross the road and rally for the goat that escapes the slaughterhouse, but we turn a blind eye every day to the atrocities that our food system engages in as a matter of course. We don't want to know what really goes on because we fear our hearts won't be able to handle the sadness, but our ignorance threatens our health, the health of our children, and the health of billions of animals.

I believe that more and more of us are going to become crusaders to take back our food system. Healthy food has become increasingly inaccessible to those without means. Corporations are taking away the ability of family farmers to enjoy full ownership and control over their farms, as they inflate the prices of machinery, seeds, fertilizers, and goods needed to operate. At the same time, processors suppress the prices that farmers are paid, leaving them with such thin profit margins that they are forced to either expand into large operations or get out of the farming business altogether. They are saddled with unmanageable debt, and forced to be complicit in unhealthy practices and inhumane treatment of animals. They are in desperate straits. Our nation is becoming sicker and sicker, and the local farmers who are trying to do right are simply unable to compete with the industrial farms.

We need to show that truly healthy food and humane animal practices matter to us. We're all busy, but there are a lot of us. If we all do a little better, a lot of good is possible.

There are organizations and people advocating for the rights of animals and standing up to the farm industry. Whether they are working to make changes in legislation, adding vegan alternatives into their diet, doing meat-less meals more often, choosing to live a vegetarian or—even better—vegan lifestyle, educating others, or caring for the animals themselves, these people are all moving us in the right direction.

If you are motivated to do your part and stand up to the industrial farming world, there are simple ways to get involved. Contact Farm Sanctuary and join its Compassionate Communities Campaign. Ask Farm Sanctuary to provide you with the literature, videos, and further information you need to spread the word about vegan or vegetarian living or to educate your own community on the horrors of factory farming and why it needs to stop. Join the ASPCA Advocacy Brigade and stay informed about farm legislation, learn how to write your own advocacy letters, and get yourself involved in the activities of the state and federal governments by signing petitions to improve a specific situation or draw attention to a specific issue. Introduce your kids and their teachers to the Jane Goodall Institute and the Roots and Shoots program, whose mission it is to advance the power of individuals to take informed and compassionate action to improve the environment of all living things.

When we make changes in our lives—whether big or small—that move us toward honoring our values and seeking compassion for others, we are moving in the right direction.

FOSTER A FARM FRIEND:

How It Works

ANIMAL SANCTUARIES SUCH as Farm Sanctuary provide a home for farm animals that have come from deplorable situations. These places are filled with loving, caring people who help to provide the rescued animals with food and water, shelter and grass, veterinary care, rehabilitation, and, at last, love and peace. When a farm sanctuary is run well, every animal is treated like the individual that it is.

Supporting a farm sanctuary gives you the option of getting involved and becoming a loving caregiver, whether from afar or up close. Your support is the perfect way to bring the animals of the farm into your family and into your heart.

Throughout the country there are hundreds of sanctuaries that are caring for farm animals. Take a look online at www.sanctuaries.org to find a sanctuary in your state. If you do not find one close to your home, I recommend visiting Farm Sanctuary's website and choosing it as your sanctuary. You'll discover there are tons of ways to get involved no matter where you call home.

If possible, schedule a day to visit the animals and get to know them up close. If that's not possible, read their stories online. You'll find that no animal is referred to by a number. Every animal has a name.

Farm Sanctuary is bursting with inspirational stories. Following are a few of my favorites.

SEND A CARE PACKAGE

As a farm sanctuary supporter, you can send a care package to help provide the sanctuary with supplies. It's the perfect way to add a farm animal to your circle of friends.

Sanctuary websites often include wish lists of items they need to help care for and treat their farm animals. As the majority of sanctuaries are not-for-profit, they rely heavily on donations of goods and dollars to gather all the equipment needed to meet their residents' needs. It's best to check with your sanctuary before sending a care package, just to make sure that they can accept your gift and put the items to good use.

- **Towels and blankets.** Towels and blankets can be new or recycled. Send only those that are smooth and will not allow claws to become caught or entangled. Soft, nonwool blankets and towels are perfect for animals to sleep and nestle on in their barns.

- **Kennel pads.** These pads are often used for dogs but are just as useful for many of the animals in sanctuaries, particularly those that are older or injured and need a little bit of extra cushioning to lie on.

- **Sunflower seeds.** These seeds are a great source of protein for chickens through the fall molting season. Cows, goats, sheep, and pigs are all able to eat and digest sunflower seeds, and the seeds' high oil content is valuable in providing energy to the animals.

- **Box or crate.** Choose a wooden box for your care package that can also double as a nesting box for the animals at the sanctuary. Make sure that your wooden box has drainage and ventilation holes.

- **Mixture of fresh fruits and veggies.** Apples, bananas, carrots, and cranberries are just some of the fruits and vegetables that farm sanctuaries collect to feed to their animals.

- **Canned fruits and vegetables.** Canned yams and canned fruits last longer than fresh foods, are easy to store, and are great for feeding to farm animals.

NIKKI THE SURVIVOR PIG

In 2008, a large flood ripped through the state of Iowa, causing much destruction to local business and roadways, and forcing many residents to evacuate their homes. One such business that was greatly affected was a large industrial farm in the area. Thousands of pigs kept in crates on the farm were put at risk when the floodwaters came crashing in, as no specific emergency escape route for the animals had been put into place. Trapped in their tiny confined crates, many helpless pigs were left to drown.

Nikki was one of the few to survive. She was released from her crate as the waters crept in and was left to fend for herself. When she finally got to dry land, she gave birth to seven piglets, which she kept safe until rescue workers from Farm Sanctuary were able to save her. Once taken in by the folks at Farm Sanctuary, Nikki showed just how loving, caring, and amazing pigs can truly be. She doted on her piglets, building them elaborate nests made out of warm straw to sleep in. And to this day, Nikki is still as protective of them as she was when they weighed only a few pounds.

Gestation sows like Nikki spend two to three years in crates and are impregnated two or three times a year. In these crates they are given no room to move and no space to turn. They do not get the chance to exercise and have no opportunity to forge relationships with their piglets once they have birthed them. For devoted mothers, this is traumatic. Though the circumstances that led to Nikki's rescue by Farm Sanctuary were terrible, she is now thankfully able to live a happy life with her piglets in a caring environment, cage free.

PATRICK THE GOAT

When Patrick arrived at Farm Sanctuary in Watkins Glen, New York, after being kept in a small pen on a Long Island property, he had deformed hooves and contorted legs and was in so much pain that he could no longer walk. At the age of one, Patrick was forced to crawl wherever he needed to go. His obvious and severe mistreatment attracted the attention of neighbors, who called the authorities. Soon after, the staff at Farm Sanctuary was informed of a goat in need and arranged to provide him refuge.

Patrick's injuries were a result of improper housing and care. His two extremely overgrown hooves were too painful for him to walk on and, as a result of crawling, the tendons in his legs had tightened, making him no longer able to walk at all. When the leaders at Farm Sanctuary met Patrick at Cornell University Hospital for Animals, he had a fever and an elevated heart rate from the pain he had endured. Doctors performed surgery to cut the tendons and applied braces to help straighten his legs. It took weeks of rest and medication, but in time, Patrick was up and walking again. With physical therapy and proper care, he is now living his life in the pastures of Farm Sanctuary, healing, playing, and no longer living his life in pain.

TRICIA AND SWEETY

There is no better example of blind love than the love that blossomed between Tricia and Sweety, two cows at New York's Farm Sanctuary. They had a lot in common even before they met: they were both dairy cows on large industrial farms; they had both borne numerous calves, each of which had been taken away from them after every birth; they were both chosen to be sent to the slaughterhouse once they were deemed to be no longer useful—and they were both blind. Tricia was the first to be rescued, taking up residence at Farm Sanctuary and befriending a cow named Linda. After Linda lost a battle to cancer, Tricia mourned for her dear friend and seemed to be at a loss. When Farm Sanctuary heard about Sweety's need for a good home, they knew she would be the perfect pick-me-up for Tricia.

Poor Sweety had spent eight years on the production line (double the usual duration) and was living in horrible conditions that rendered her lame. After she gave birth to a set of twins—which is looked down upon in the dairy world, as the males are usually small and the females typically sterile— Sweety's "value" dropped considerably, she was considered "spent" and was about to be sent to slaughter. Thankfully, a woman from a local horse rescue heard about the blind cow and asked if instead she could find her a good home, which she did at Farm Sanctuary. When Sweety arrived at Farm Sanctuary's gates, Tricia could smell her presence and lit up immediately. Sweety was led into Tricia's stall, where they took some time to sniff each other out. They spent the day chewing hay together, walking beside each other in the paddock, and becoming fast friends.

In an industrial farm and along the slaughter line, blind cows that have a heightened sense of hearing can become especially terrified and overwhelmed by the banging of metal gates, the clanking of shackles, and the bellowing of their herd mates. This type of torture is scarring and not something that can easily be forgotten, but thanks to the good people at Farm Sanctuary, Tricia and Sweety now live peaceful lives out in the green pastures, with little to make them anxious. They are surrounded by people who love and care for them, and are doing what cows do best: making friends.

SPONSOR A FARM ANIMAL

You don't have to own a farm in order to care for a farm animal. Most farm sanctuaries offer the opportunity for members of the public to sponsor or "adopt" a farm animal. This process involves choosing an animal at the sanctuary that you truly love and have a desire to help, and donating your money to help provide that pig, cow, chicken, sheep, goat, or duck with the food, shelter, and care it needs to live a happy and healthy life. When you adopt an animal through Farm Sanctuary, you even get an adoption certificate, photographs of your farm friend, and top priority to pay him or her a visit at the sanctuary as often as you like. Ask the leaders at your local sanctuary about their sponsorship programs and how you can become the proud parent of a beautiful farm animal today.

STEWART FAMILY TRADITIONS:

Sponsoring a Farm Animal on Thanksgiving

I MAY HAVE EXAGGERATED earlier when I mentioned I wasn't a very good Thanksgiving guest. In reality, I try very hard not to be insufferable or preachy on this day. Our Thanksgiving table welcomes vegans, vegetarians, and meataholics. In keeping with the holiday's theme of giving thanks, I make a donation in the name of each of my guests through Farm Sanctuary's Adopt a Farm Animal Program. Rather than finding a place card at their seats, my guests each find an envelope that contains a short profile and a beautiful photo of the guest's animal adoptee, courtesy of Farm Sanctuary. The guests are always excited to see which animal they'll get, whether it's Ingrid the goat, Elfie the duck, Nate the pig, Bear the sheep, Daisy the turkey, or Dandypants the rooster.

I only ask that my guests be animal lovers. As animal lovers, we can always do and be more for those creatures who are depending on us. I am most grateful to be surrounded by loving, compassionate, and giving people, and I am constantly reminded of how many people and animals don't have this good fortune. Let's do our best to find them and introduce some of this good fortune into their lives as well.

Acknowledgments

Other than myself, three people needed to be born for this book to be.

First, Lia Ronnen. Lia convinced me that what I thought was the insane way I lived my life was actually worth sharing and, dare she say, inspirational enough to fill a book. Lia is a true visionary, and her devotion to animal advocacy kept what others might have been afraid to print in this book. She is my fairy godmother.

The second was Rachel Filler. Rachel flew above this whole project, organizing and holding all the content in her heart. Luckily for me, her brain and heart are enormous. She is my rock.

Then Lisel Ashlock brought lush life to all our precious friends. She went beyond the job of illustrator, along with Lia and Rachel conceptualizing, designing, and building this book. Lisel is my dream weaver.

I also want to give tremendous thanks to . . .

Te Chao, Mura Dominko, Michelle Ishay-Cohen, Sibylle Kazeroid, Paul Kepple, Nancy Murray, Toni Poynter, and Kara Strubel for being part of the team.

Jodi Levine for her inspired crafts and ingenuity. Her enthusiasm for the project was part of its fuel.

Purl Soho for providing the most delicious craft materials.

Andrea Arden, Joanne Basinger, and Mike Lustig for consulting on this book and, more important, consulting on my life.

Vyolet and Drayton Michaels for consulting on the book and also for inspiring me to do better for my animals every day. Vyolet is my true accomplice. Little did she know when she signed up to train my dogs that she'd get sucked into pig socialization, bunny-bonding, parrot enrichment, horse clicker-training… Drayton is an invaluable teacher. I have yet to hear anyone speak as eloquently, intelligently and reasonably about pit bulls.

Animal Haven Shelter and the Humane Society of New York for allowing my family to become part of theirs.

Gene Baur for starting me on the path of living mindfully. Susie Coston for being the woman I hope to become. And Sylvia Moskovitz for being the woman I want to be adopted by.

Karen Jolsen for her advice and support.

Erin O'Sullivan for her senior dog expertise and for being born with a fire in her soul to help.

Ray LaMontagne for singing to me and my animal family every day while I wrote.

Sophie Gammand and Elias Freeman for allowing us to illustrate some of their masterpieces.

My awesome animal-loving parents. Had they not had me, my brother, and my sister back-to-back, my guess is our home would have looked a lot like mine today. They filled me up with love so that I'd have lots to give.

Resources

American Society for the Prevention of Cruelty to Animals (ASPCA)
www.aspca.org

Andrea Arden Dog Training
www.andreaarden.com

Animal Haven Shelter
www.animalhavenshelter.org

Animal Legal Defense Fund
www.aldf.org

Animal Place (directory of farm animal sanctuaries)
www.sanctuaries.org

Animal Welfare Approved
animalwelfareapproved.org

Drayton Michaels, CTC
www.pitbullguru.com
www.urbandawgs.com

Farm Sanctuary
www.farmsanctuary.org

The Great Backyard Bird Count
http://gbbc.birdcount.org/

H.E.A.R.T.
www.teachhumane.org

The Jane Goodall Institute
www.janegoodall.org

Jane Goodall's Roots and Shoots
www.rootsandshoots.org

Jodi Levine
www.supermakeit.com

Mamma Biscuit
www.mammabiscuit.com

National Wildlife Rehabilitators Association (NWRA)
www.nwrawildlife.org

Petfinder
www.petfinder.com

Susie's Senior Dogs
www.susiesseniordogs.com

Vyolet Michaels, CTC and CPDT-KA
www.urbandawgs.com

Wildlife Rehabber
www.wildliferehabber.com

Index